A Theory of Syntax

Human language seems to have arisen roughly within the last 50–100,000 years. In evolutionary terms, this is the mere blink of an eye. If this is correct, then much of what we consider distinctive to language must in fact involve operations available in pre-linguistic cognitive domains. In this book Norbert Hornstein, one of the most influential linguists working on syntax, discusses a topical set of issues in syntactic theory, including a number of original proposals at the cutting edge of research in this area. He provides a theory of the basic grammatical operations and suggests that there is only one that is distinctive to language. If this theory is correct then this narrows the evolutionary gap between verbal and non-verbal primates thus facilitating the rapid evolutionary emergence of our linguistic capacity.

NORBERT HORNSTEIN is Professor in the Department of Linguistics at the University of Maryland, College Park. He has written several books on minimalist syntax including *Understanding Minimalism* (with J. Nunes and K. Grohmann, 2005) and *Move!: A Minimalist Theory of Construal* (2000).

A Theory of Syntax

Minimal Operations and Universal Grammar

Norbert Hornstein

University of Maryland

CAMBRIDGE
UNIVERSITY PRESS

CAMBRIDGE UNIVERSITY PRESS
Cambridge, New York, Melbourne, Madrid, Cape Town, Singapore, São Paulo, Delhi, Dubai, Tokyo

Cambridge University Press
The Edinburgh Building, Cambridge CB2 8RU, UK

Published in the United States of America by Cambridge University Press, New York

www.cambridge.org
Information on this title: www.cambridge.org/9780521728812

First published 2009
Reprinted with corrections 2009

Printed in the United Kingdom at the University Press, Cambridge

A catalogue record for this publication is available from the British Library

Library of Congress Cataloguing in Publication data
Hornstein, Norbert.
A theory of syntax : minimal operations and universal grammar / Norbert Hornstein.
p. cm.
Includes bibliographical references and index.
ISBN 978-0-521-44970-0 (hardcover) – ISBN 978-0-521-72881-2 (softcover)
1. Grammar, Comparative and general – Syntax. 2. Minimalist theory (Linguistics)
3. Language and languages – Origin. 4. Evolution. I. Title.
P291.H662 2009
415 – dc22 2008033468

ISBN 978-0-521-44970-0 Hardback
ISBN 978-0-521-72881-2 Paperback

Contents

Preface

Books are to insights what belatedly closed barn doors are to horses. By the time they get finished, it is not entirely clear (at least to the author) why you wrote them and why it all took so long. This particular project has some immodest aims. Here are the two central ones.

First, it tries to outline (yet again) a way of understanding the minimalist project. This time around, I try to provide a rarefied empirical motivation. Following the lead of Hauser, Chomsky and Fitch (2002) I trot out an evolutionary argument called, unoriginally, "Darwin's Problem."[1] I couple this with a second neurobiological reason based on Poeppel and Embick (2005) which, following them, I call the Granularity Mismatch Problem. These two problems, I propose, should function as high-level empirical boundary conditions on adequate accounts of the properties of Universal Grammar (UG) and the structure of the Faculty of Language (FL), much as Plato's Problem has in earlier inquiry. Thus, theories of UG and FL will have to address all three problems to be explanatorily adequate. The addition of this pair of requirements on explanatory adequacy is the central contribution of the Minimalist Program.

Second, it outlines a way of operationalizing these concerns by proposing a particular theoretical project: to derive the properties of UG from simpler, more natural empirical primitives. This project is very like the one outlined in Chomsky (1977) with regard to Ross's islands. Both begin from the assumption that earlier accounts are roughly empirically correct. Thus, Chomsky (1977) assumed that Ross's (1967) constraints were more or less empirically adequate and wanted to "explain[ed them] in terms of general and quite reasonable 'computational' properties of formal grammar" (p. 89). So too we will here assume that Government Binding Theory (GB) correctly limns the properties of UG/FL and our aim is to explain them on the basis of simpler, more general, more natural cognitive operations and principles. The effort requires moving from general programmatic desiderata to particular theoretical proposals, i.e. from Minimalist Program to Minimalist Theory. The core of the present proposal is

[1] I am sure that Chomsky is responsible for this term. However, I have not been able to track down where it was first introduced. Cedric Boeckx has used this term in Boeckx (forthcoming).

a theory of basic operations, one of which is unique to language (viz. Label). The aim is to show how the general features of FL might be derived from this inventory. The basic idea is that Label together with the other basic operations (Concatenate, Copy) plus a computational principle which requires minimizing dependency length suffice to yield a system with many of the properties of a GB style account.[2] The chief novelty of the proposal involves a reinterpretation of Minimality in terms of Paths and a particular understanding of labeling. Labeling functions to "close" concatenation in the domain of the lexical items (LI). As a result it creates equivalence classes of objects grounded in each LI. By closing concatenation in the domain of the LIs, hierarchy emerges. By creating equivalence classes, constituency arises. That grammatical operations target constituents follows from how Concatenate is restricted to LIs and their labeled "equivalents." Thus, three of the central features of natural language grammars emerge as by-products of labeling.

This is the basic proposal. The details are what take up seven chapters.

One last word before plunging in; most books are social constructions. They live in a rich eco-system populated by the research of others and, further, require the support and indulgence of many colleagues to grow. This is especially so for this one. I have many intellectual debts. Most prominently, the project is inconceivable in the absence of Chomsky (1995a) and the subsequent minimalist papers, especially Hauser *et al.* (2002) and Chomsky (2005a). Though I differ in detail with many of Chomsky's later minimalist proposals, I have found the general problem he outlined to be endlessly stimulating and have also found that the contours of my own views emerged most clearly when backlit by these later minimalist proposals.

The style and substance of the present project has also been greatly influenced by Boeckx (2008). Boeckx's work is the most carefully thought out version of an Agree-based minimalism that I am acquainted with. Given my skepticism concerning such approaches, it has been extremely helpful to have Boeckx's views (as well as Cedric himself) to consult.

To an equal degree, the ideas contained here reflect ones contained in a forthcoming book by Paul Pietroski on basic operations in semantics. This book has heavily borrowed from his. Being able to talk to Paul and read his stuff has been invaluable and this project would have seriously floundered without his generous indulgence. He is the Platonic form of the colleague.

Let me also thank Juan Uriagereka. Since 1993, we have carried on a spirited conversation about Minimalism. We have argued about the aims of the program, the basic theoretical concepts to develop and the best techniques for their

[2] I say "GB style" for I include in this GB's cousins including LFG, GPSG, HPSG and RG. Though the particulars of GB are what I concentrate on, all the above mentioned approaches cut grammars along more or less the same joints.

implementation. We have agreed, disagreed, reagreed and even misagreed over issues large and small. From all of this I have learned immeasurably.

Last of all, Chametzky (1996) and Epstein (1999) have heavily influenced the ambitions of the present project. Both are unabashed theoretical works whose aim is to elucidate and polish the basic concepts of our discipline. All too often such work is disparaged as non-empirical. This is unfortunate. There are many roads to insight. One of these faces inwards to the basic concepts rather than outward to empirical consequence. There is value in outlining how basic ideas fit together independently of whether they have empirical utility. This kind of theoretical enterprise, I believe, is of particular value right now and is central to the minimalist enterprise. Of course, like all potentially valuable pursuits, it carries its own risks. But this is a very bad reason not to pursue its potential rewards.

Many people have discussed the issues contained in what follows with me at length. Only those who have had the misfortune of having me descend upon them with an *idée fixe* can truly appreciate how much this puts me in their debt. I would like to specifically mention Cedric Boeckx, Željko Bošković, Rob Chametzky, Sam Epstein, Tim Hunter, Bill Idsardi, Jairo Nunes, Paul Pietroski, David Poeppel, Juan Uriagereka and Matt Wagers.

Last of all, special thanks to Cedric Boeckx, Jairo Nunes and Paul Pietroski for comments on an earlier draft, endless interminable discussion of half-baked ideas and well-placed skepticism that I have only occasionally taken to heart. Also, special thanks to Akira Omaki for his hard work in getting the MS ready for publication.

A final point, on going to press I discovered that Robert Berwick (1997: Syntax facit saltum, *Journal of Neurolinguistics*, 10:2/3, 231–249) had earlier developed the general line of attack pursued here.

1 Minimalism and Darwin's Problem

1.1 Introduction

Contemporary generative theorists are united by (at least) one conviction and divided by (at least) one other. What unites everyone is the understanding that grammatical knowledge is rule based. Native speakers of a given language L have mastered rules for L that allows them to generate an unbounded number of tokens of L (i.e. sentences, phrases, etc.). Rules are required because the tokens of L are for all practical purposes infinite and thus cannot possibly be stored individually in a finite organism. The rule-based character of linguistic knowledge is, thus, not controversial among generative grammarians.[1]

What is controversial is how these grammars are structured; what kinds of rules they allow, what kinds of primitive relations they exploit and what kinds of elements they involve. Here there is a lot of controversy. One line of inquiry with which I am very partial, the Minimalist Program, takes it as a boundary condition on inquiry that the basic operations of UG be simple and that the attested complexities of natural language be the result of the interactions of simple subsystems. This vision gains teeth when the meaning of "simple" is filled out. Here is how I understand the term.

There are several dimensions to simplicity.

First, simple systems are non-redundant. Redundancy arises in grammars when different operations can independently generate the same structural relations or different principles independently exclude them. An example (which is developed in more detail in later chapters) can serve to illustrate my meaning. Many current grammars postulate both a Move operation and an AGREE operation capable of operating over long distances.[2] Both serve to relate remote

[1] Which does not mean to say that it is not still controversial. There are many in the connectionist world who appear to deny the rule-based nature of grammatical knowledge. Such dissenters are happy enough to concede that natural language objects display patterns, but patterns are not rules. The problem with this view, I believe, is that it is quite clear that the number of possible *patterns* is likewise unbounded and that only rules will do. For discussion of this basic point see Jackendoff (1994).

[2] Note the qualification. That grammars involve agreement, i.e. some form of feature checking, is clear. The interesting operation is the non-local version of AGREE as it covers much the same

elements (non-sisters) to each other. All things being equal, grammars should not contain both kinds of operations as they can cover a great deal of the same empirical territory. This is not a good thing for at least two reasons. First, a UG with multiple routes to the same end gains an undesired flexibility, which adversely affects its explanatory potential. Methodologically speaking more brittle theories are more easily falsified and thus preferable. Further, more brittle UGs restrict the learner's options more than more flexible ones do. If there are two ways of covering the same data set, then the learner must choose between them, seldom a good thing given the logic of Plato's problem. Of course, things may not be equal and both operations might be required, but a good working hypothesis is that grammars are not redundant in this way.[3]

Second, in simple theories of UG the basic operations are as sparse as possible. Fewer is better. Ockham is right. All things being equal theories that employ a sparser inventory of principles and basic operations are better than those with an ampler armamentarium.[4] Of course, oftentimes things are not equal. In such cases, I am inclined to a somewhat stronger allegiance to Ockham. It is a truism that the richer a theory's apparatus the wider its empirical coverage. This means that sparser theories are expected to face empirical challenges that more ample theories will avoid. I understand this truism to mean that the latter should face more stringent explanatory demands before winning the day. Precisely because their data coverage is expected to be wider, more ample theories should either cover a hell of a lot more territory than their more restrained competitors or should do so in such ways that do not sacrifice explanatory insight. My version of Ockham strongly prefers the leaner meaner account and requires substantial advantages before it is abandoned!

Third, in simple accounts the basic operations and principles are natural. Just what makes such operations and principles "natural" is a subtle question.

territory as movement, an operation that relates remote elements to one another. See Chapter 6 for a full discussion.

3 This form of argumentation originates in GB era analyses where it was argued that principles of UG should not overlap in their domains of application. For example, Chomsky (1981: 12–14) where he notes that the fecundity of "explor[ing] redundancies in grammatical theory, that is, cases in which phenomena are 'overdetermined' by a given theory in the sense that distinct principles (or systems of principles) suffice to account for them." See also Chomsky (2005a: 10) where he notes:

> It has also proven useful over the years to ask whether apparent redundancy of principles is real, or indicates some error of analysis. A well-known example is passive forms of exceptional case marking constructions, which seemed at one time to be generated by both passive and raising transformations. Dissatisfaction with such overlapping conditions led to the realization that transformations did not exist: rather just a general rule of movement...

4 There is a good reason for this. Given that theories meet evidence "as a corporate body" (as Quine says) then the fewer the basic assumptions required to account for the evidence the greater the evidence in favor of each assumption. Hence the methodological privilege of fewer assumptions, all things being equal.

However, this has not prevented generative grammarians from arguing for and against proposals in just such terms over the years. For example, to the degree grammars facilitate "computation" they are natural, e.g. locality conditions (like subjacency or minimality) are "nice" properties from a computational point of view given the burden that distance imposes on computational efficiency and memory.[5] Another example; feature checking and copying are natural computational operations for the faculty of language (FL) to exploit as they are almost certainly operative in other cognitive domains, albeit with different expressions being copied and different features being checked. Given the rather late emergence of FL in humans it is evolutionarily natural that FL should import operations from other parts of the cognitive system. This suggests one more mark of "naturalness," namely generality; operations and principles at work in other parts of the cognitive economy are natural resources for linguistic computations. A further mark of the "natural" is the "atomicity" of the computational operations. Merge (join two expressions) and copy (duplicate an expression) are reasonably taken as computationally "atomic" operations.[6] They contrast with more complex language specific rules like "passive" which are reasonably analyzed as compiled combinations of more basic operations. This conception of "simple" and "atomic" casts a furtive glance towards implementation in brain like material. Whatever operations grammarians propose must ultimately be embedded in brain circuitry. It is reasonably clear how one could build a merge or copy circuit, and this is one reason that primitive operations like these are attractive.

I would like to stress this last point. David Poeppel and colleagues have recently emphasized that any grammatical process we propose must be embodied in brain circuitry if it is really operative in *our* FL. However, the linking hypotheses between language and brain are "most likely to bear fruit if they make use of computational analyses that appeal to *generic* [my emphasis, NH] computational subroutines" (Poeppel and Monahan in press). Thus, keeping basic operations simple and generic comes with the advantage of conceivably being implementable.[7]

In sum, FL will be natural if it is based on principles and operations that promote computational tractability, that are built from parts that are cognitively general and atomic, and that are basic enough to be (plausibly) embodied in neural circuitry.

[5] See Chomsky (1977) for discussion along these lines for subjacency. See too Berwick and Weinberg (1984).

[6] I would be inclined to say that they are primitively recursive, the building blocks for possibly more complex combinations. For discussion, see Chapter 7.

[7] For some further discussion of how primitives of grammar should relate to primitives of neuroscience, see Embick and Poeppel (2005a).

As should be evident, even given the desiderata above, there remains plenty of room for diverging views on how to interpret these guidelines and, not surprisingly, there is a large pool of potential candidates for the inventory of basic operations and principles. Nonetheless, I believe that these guidelines can play a more than rhetorical role in the construction and evaluation of grammatical proposals. More concretely, I believe that the search for simple operations and principles suggests an interesting minimalist project: the construction of grammatical models based on a small inventory of operations and principles that are at once evolutionary and neurologically plausible and from which the basic properties of natural language grammars can be qualitatively derived. The reason for this is best articulated in an evolutionary idiom.

1.2 Minimalism and Darwin's Problem

Over the last 50 years of research generative grammarians have discovered many distinctive properties of natural language grammars (NLG). For example: (a) NLGs are recursive, viz. their products (sentences and phrases) are unbounded in size and made up of elements that can recur repeatedly; (b) NLGs generate phrases which display a very specific kind of hierarchical organization (viz. that described by X′ theory); (c) NLGs display non-local dependencies (as in Wh-movement, agreement with the inverted subject in existential constructions, or reflexive binding), which are subject to hierarchical restrictions (e.g. binding relations are subject to a c-command requirement) and locality restrictions (e.g. controllers are subject to the minimal distance requirements and anaphors must be bound within local domains). These properties, among others, are universal characteristics of natural language and thus reasonably construed as universal features of human grammars. A widely adopted (and to my mind very reasonable) hypothesis is that these characteristics follow from the basic organization of FL, i.e. they derive from the principles of UG.

Given this, consider a second fact about FL: it is of recent evolutionary vintage. A common assumption is that language arose in humans in roughly the last 50,000–100,000 years. This is very rapid in evolutionary terms. It suggests the following picture: FL is the product of (at most) one (or two) evolutionary innovations which, when combined with the cognitive resources available before the changes that led to language, delivers FL. This picture, in turn, prompts the following research program: to describe the pre-linguistic cognitive structures that yield UG's distinctive properties when combined with the one (or two) specifically linguistic features of FL. The next three chapters try to outline a version of this general conception.[8]

[8] This clearly echoes the program outlined in Hauser, Chomsky and Fitch (2002).

The approach, I believe, commits hostages to a specific conception of FL. It does *not* have a high degree of internal modularity. The reason for this is that modular theories of UG suppose that FL is intricately structured. It has many distinct components that interact in complex ways. On the assumption that complexity requires natural selection and that natural selection requires time to work its magic (and lots of it: say on the order of (at least) millions of years), the rapid rise of language in humans does not allow for this kind of complexity to develop.[9] This suggests that the highly modular structure of GB style theories should be reconsidered.

Fodor (1998) puts the logic nicely:

> If the mind is mostly a collection of innate modules, then pretty clearly it must have evolved gradually, under selection pressure. That's because . . . modules contain lots of specialized information about problem-domains that they compute in. And it really would be a miracle if all those details got into brains via a relatively small, fortuitous alteration of the neurology. To put it the other way around, if adaptationism isn't true in psychology, it must be that what makes our minds so clever is something pretty general . . .

What holds for the modularity of the mind holds for the modularity of FL as well.[10] A highly modular FL has the sort of complexity that requires adaptation through natural selection to emerge. In addition, adaptation via natural selection takes lots of time. If there is not enough time for natural selection to operate (and 50,000–100,000 years is the blink of an evolutionary eye), then there cannot be adaptation, nor this kind of highly modular complexity. The conclusion, as Fodor notes, is that the system of interest, be it the mind or FL, must be simpler and more general than generally thought.

Lest I be misunderstood, let me make two points immediately.

First, this reasoning, even if sound (and it is important to appreciate how speculative it is given how little we know about such evolutionary matters in the domain of language) does not call into question the idea that FL is a distinct cognitive faculty. What is at issue is not whether FL is modular with respect to other brain faculties. Rather what we are questioning is the internal modular organization of FL itself. The standard view inherited from GB (and

9 The assumption that complexity requires natural selection is a standard assumption. For example, Cosmides and Tooby (1992), Dawkins (1996) and Pinker (1997) quoted in Fodor (2000: 87). Dawkins's words serve to illustrate the general position:

> whenever in nature there is a sufficiently powerful illusion of good design for some purpose, natural selection is the only known mechanism that can account for it. (p. 202)

10 Fodor (2000) might not accept this inference as he takes the program in linguistics to only be interested in knowledge not mental mechanisms. I am inclined to think that Fodor is incorrect in his characterization of Chomsky's position. However, what is relevant here is that grammars are construed as interested in the mechanics of linguistic mentation. The inventory of rules and principles describe real mechanisms of the mind/brain.

I believe still with us today) is that FL itself is composed of many interacting grammatical subsystems with their own organizing principles. For example, the Binding Theory has its proprietary locality conditions (i.e. Binding Domains), its own licensing conditions (i.e. Principles A, B and C), and its own special domain of application (i.e. reflexives, pronouns and R-expressions). So too for Control, Case Theory, Theta Theory, etc. It is this kind of modularity that is suspect as it requires FL to have developed a lot of complicated structure in a rather short period of time both internal to FL itself and internal to each module of FL. If this is not possible because of time constraints, then rich internal modularity is not a property of FL.

Second, I assume that the generalizations and "laws of grammar" that GB discovered are roughly empirically correct. In my opinion, one of the contributions of modern generative grammar to the study of language has been the discovery of the kinds of properties encapsulated in GB.[11] Reconsidering the internal modular structure of GB does not imply rejecting these generalizations. Rather it takes as its research goal to show that these generalizations are the products of more primitive factors. The proposal is to add to the agenda of grammatical theory the aim of deducing these "laws" from more basic principles and primitives.[12]

A picture might be of service here to get the main point across.

(1) Pre-linguistic principles and operations → ?? → (roughly) GB laws

This picture is intended to invoke the more famous one in (2).

(2) Primary Linguistic Data (of L)→ UG → Grammar (of L)

The well-known picture in (2) takes the structure of FL as a black box problem, dubbed "Plato's Problem" or the logical problem of language acquisition. The goal is to study what UG looks like by constructing systems of principles that can bridge the gap between particular bits of PLD to language particular grammars consistent with that PLD. Generativists discovered that the distance between the two is quite substantial (as the information provided by the PLD significantly underdetermines the properties of the final state of FL) and so

11 The generalizations characteristic of GB have analogues in other generative frameworks such as LFG, GPSG, Tag Grammars, Relational Grammar etc. In fact, I consider it likely that these "frameworks" are notational variants of one another. See Stabler (2007) for some discussion of the inter-translatability of many of these alternatives.

12 There is a term in the physical sciences for the status I propose for GB. The roughly correct theory whose properties are targets for explanation is called an "effective theory." Being an "effective theory" is already a mark of distinction for to be one, a theory must have good empirical credentials. However, the term also implies that the structural properties of an effective theory need further elucidation and which will come from being subsumed in a more general account. As such, treating GB (and its analogues, cf. note 11) as an effective theory is to at once praise its accomplishments and ask for more theoretical refinement.

requires considerable innate mental structure (including the principles of UG) to bridge the gap. GB is one well-articulated proposal for the structure of UG that meets this "poverty of stimulus" concern.

An important feature of the GB model is its intricate internal modularity as well as the linguistically dedicated aspects of its rules and principles. The modules in a GB system are specifically linguistic. By this I mean that their structures reflect the fine details of the linguistic domains that concern them rather than being reflections of more general cognitive mechanisms applied to the specific problems of language.[13] On this conception, FL is a linguistically dedicated system whose basic properties mirror the fine structures of problems peculiar to language; problems related to antecedence, binding, displacement, agreement, case, endocentricity, c-command etc. These latter are specifically linguistic in that they have no obvious analogues in other cognitive domains. It is fair to say that GB is cognitively exceptional in that its principles and operations are cognitively *sui generis* and very specific to language.[14] In other words, GB endorses the view that FL is cognitively distinctive in that its internal structure displays few analogues with the principles and operations of other cognitive modules. In Chomsky's (2005a) terminology, GB reflects the view that linguistic competence is replete with first factor kinds of ingredients and that third factor processes are relatively marginal to explaining how it operates.

The picture in (1) is modeled on that in (2). It proposes taking the reasoning deployed in (2) one step further. It relies on the belief that there is an analogy between learning and evolution. In both cases development is partially a function of the environmental input. In both cases it is also partially a function of the prior structure of the developing organism. In both cases the "shaping" effects of the environment on the developmental processes requires reasonable

[13] Fodor (1998) characterizes a module as follows:

> A module is a more or less autonomous, special purpose, computational system. It's built to solve a very restricted set of problems, and the information it can use to solve them with is proprietary.

This is a good characterization of GB modules. They are autonomous (e.g. to compute case assignment one can ignore theta roles and similarly licensing binding relations can ignore case and theta properties) and special purpose (e.g. case vs. theta vs. binding). The problems each addresses are very restricted and the concepts proprietary (e.g. binding, control).

[14] As Embick and Poeppel (2005a) observe, this is a serious problem for those aiming to find brain correlates for the primitives of FL. They dub this the granularity problem. They propose that one aim of linguistics and neuroscience should be to solve this problem by finding a level that can serve to relate the basic conceptions of each. Their concrete proposal is that an appropriate level of abstraction is the "circuit." Circuits are brain structures that compute simple operations. The aim is to find those primitive operations that are at once empirically grounded and that could be embodied in neural wet-ware. Given this, the goal for the minimalist will be to find a class of very basic primitive operations that plausibly underlie linguistic computations for consideration as candidates for possible neural circuits.

time during which the environment can "shape" the structures that develop.[15] (1) takes the evolution of the principles of UG as a function of the pre-linguistic mental state of "humans" and something else ("??"). Moreover, we know whatever "??" is, it must be pretty slight – a new kind of operation or principle – given that FL/UG emerged quite rapidly. We can investigate this process abstractly (let's call it the logical problem of language evolution or "Darwin's Problem") by considering the following question: *what must be added to the inventory of pre-linguistic cognitive operations and principles to deduce the principles of UG?*[16] We know that whatever is added, though pretty meager, must be sufficient when combined with the resources of non-specifically linguistic cognition to derive a system with the properties summarized by GB. In other words, what we want is an operation (or two) that once added to more general cognitive resources allows what we know about FL to drop out. On this conception, what is specifically linguistic about FL's operations and principles is actually rather slight. This is in strong contrast to the underlying ethos of GB, as noted above. The logic of Darwin's Problem argues against the cognitive exceptionalism of FL. Its basic operations and principles must be largely recruited from those that were pre-linguistically available and that regulate cognition (or computation) in general. FL evolved by packaging these into UG and adding one novel ingredient (or two). This is what the short time frame requires. What (1) assumes is that even a slight addition can be very potent given the right background conditions. The trick is to find some reasonable background operations and principles and a suitable "innovation."

Once again, the sense of the program is well expressed in Fodor (1998):

> . . . it's common ground that the evolution of our behavior was mediated by the evolution of our brain. So what matters with regard to the question whether the mind is an adaptation is not how complex our behavior is, but how much you would have to change an ape's brain to produce the cognitive structure of the human mind . . . Unlike our minds, our brains are, by any gross measure, very like those of apes. So, it looks as though small alterations of brain structure must have produced very large behavior discontinuities from the ancestral apes to us.

This applies to the emergence of linguistic facility as well, surely the most distinctive behavioral difference between us and our ape ancestors.

Note two more points: First, evolutionary explanations of behavior, as Fodor rightly insists, piggy-back on changes in brain structure. This is why we would like our descriptions to be descriptions (even if abstract) of mechanisms and

[15] These analogies between learning and evolution have long been recognized. For an early discussion in the context of generative grammar, see Chomsky (1959). As Chomsky's review makes clear, the analogy between learning and evolution was recognized by Skinner and was a central motivation for his psychological conceptions.

[16] The term "Darwin's Problem" is taken from Boeckx (forthcoming).

processes plausibly embodied in brains (see note 14). Second, as Fodor correctly observes, much of this talk is speculative for very little (Fodor thinks "exactly nothing") is known of how behavior, linguistic or otherwise, supervenes on brain structure. In the domain of language, we know something about how linguistic competence relies on grammatical structure and one aim of the Minimalist Program as I understand it is to investigate how properties of grammars might supervene on more primitive operations and principles that plausibly describe the computational circuitry and wiring that the brain embodies.

Many minimalist proposals can be understood as addressing how to flesh (1) out. Chomsky (2005a) is the prime text for this. As he notes, there are three kinds of principles at work in any specific grammar: (i) the genetic endowment (specific to language), (ii) experience, and (iii) principles that are language or even organism independent. Moreover, the more that any of these can explain a property of grammar, the less explanatory work needs to be done by the others. What modern generative grammar has investigated is the gap between experience and attained linguistic competence. What minimalism is studying is the gap between the third factor noted above (non-specifically linguistic principles and operations) and the first factor (what UG needs that is not already supplied by third factor principles). The short evolutionary time scale, Chomsky (2005a: 3) suggests, implicates a substantial role for principles of the third kind (as do Fodor's 1998 speculations noted above). The inchoate proposal in (1) is that this problem is fruitfully studied by taking the generalizations unearthed by GB (and its cognates, cf. note 11) as the targets of explanation (i.e. by treating GB as an effective theory).

Before moving on, I would like to emphasize one more point.[17] As conceived here, the Minimalist Program is clearly continuous with its GB predecessor in roughly the way that Darwin's Problem rounds out Plato's. GB "solves" Plato's problem in the domain of language by postulating a rich, highly articulated, linguistically specific set of innate principles. If successful, it explains how it is that children are able to acquire their native languages despite the poverty of the linguistic input.[18] This kind of answer clearly presupposes that the sorts of mechanisms that GB proposes could have developed in humans. One source of skepticism regarding the generative enterprise is that the structures that UG requires if something like GB is correct could simply not have arisen by standard evolutionary means (e.g. by natural selection given the short time period involved). But if it could not have arisen, then clearly human linguistic facility cannot be explained by invoking such mechanisms. Minimalism takes

[17] This addition owes a lot to discussions with Paul Pietroski.

[18] As the reader no doubt knows, this overstates the case. Principles and Parameters accounts like GB have not yet accounted for how children acquire language. The problem of how parameters are set, for example, is very difficult and as yet unresolved. See Chapter 7 for some additional discussion.

this concern to heart. It supposes that FL could arise in humans either by the shaping effects of experience (i.e. through natural selection) or as a by-product of something else, e.g. the addition of new mechanisms to those already extant. For natural selection to operate requires considerable amounts of time. As it appears that FL emerged recently and rapidly as measured in evolutionary time, the first possibility seems to be ruled out. This leaves the "by-product" hypothesis. But a by-product of what? The short time scale suggests that the linguistic specificity of FL as envisaged by GB must be a mirage. FL must be the combination of operations and principles scavenged from cognition and computation in general *with possibly small adventitious additions.* In other words, despite appearances FL is "almost" the application of general cognitive mechanisms to the problem of language. The "almost" signals the one or two innovations that the 50,000–100,000 year time frame permits. The minimalist hypothesis is that FL is what one gets after adding just a little bit, a new circuit or two, to general principles of cognition and computation. If this is "all" that is distinctive about FL it explains how FL could have rapidly emerged in the species (at least in embryonic form) without the shaping effects of natural selection. The Minimalist project is to flesh this picture out in more concrete terms.[19]

1.3 Two more specific minimalist research projects

To advance this theoretical goal two kinds of projects are currently germane. The first adopts a reductive strategy. Its goal is to reduce the internal modularity of UG by reducing apparently different phenomena to the same operations. This continues the earlier GB efforts of eliminating "constructions" as grammatical primitives by factoring them into their more primitive component parts.[20] Two examples will illustrate the intent.

An important example of reduction is Chomsky's (1977) proposal in "On wh-movement." Here Chomsky proposes unifying the various kinds of constructions that display island effects by factoring out a common movement operation involved in each. In particular, Wh-movement, Topicalization, focus-movement, tough-constructions, comparative-formation and Relativization all display island effects in virtue of involving Wh- (or later, A′-) movement

[19] This way of stating matters does not settle what the mechanism of evolution is. It is compatible with this view that natural selection operated to "select" the one or two innovations that underlie FL. It is also compatible with the position that the distinctive features of FL were not selected for but simply arose (say by random mutation, or as by-products of brain growth). This is not outlandish if what we are talking about is the emergence of one new circuit rather than a highly structured internally modular FL. Of course, once it "emerged" the enormous utility of FL would insure its preservation through natural selection.

[20] See Chomsky (1983) for discussion.

subject to subjacency. What heretofore were treated as different kinds of constructions, are henceforth treated as involving a common core operation (Wh/A′ movement) subject to a common condition (subjacency). The island effects the disparate constructions display are traced to their all having Wh/A′-movement as a key component. In other words, sensitivity to island conditions is a property of a particular construction in virtue of having Wh/A′ movement as a sub-component.

This reduction of island sensitive constructions to those involving Wh/A′ movement as a subpart was not taken to imply that, for example, Topicalization and Relativization were *identical* constructions. Their distinctive features were and are obvious. However, despite their differences because all these "constructions" use the same basic Wh/A′-movement operation they will all be subject to the subjacency condition and so all display island effects when this condition is violated. Thus, the island characteristics of these various constructions are explained by analyzing each as involving a common building block, the operation of Wh/A′-movement. Why do Topicalization and Relativization and Question formation etc. all obey island conditions? Because whatever their other differences, they all involve the operation of Wh/A′-movement and Wh/A′ movement is subject to subjacency.[21]

A second example of this kind of reductive reasoning is pursued in Hornstein (2001). It attempts to reduce obligatory control and principle A to conditions on movement. More generally, the proposal is that all feature checking occurs under Merge, that Move involves an instance of Merge (viz. it is the complex of Copy and Merge) and that merging into multiple thematic positions via Move is possible. This has the effect of reducing obligatory control and principle A to the theory of movement (along with case theory, as first proposed in Chomsky 1993), which, in turn, reduces the modularity of UG by reducing case, theta and antecedence relations to those constructible via licit applications of Merge and Move. This can be construed as a version of the Chomsky (1977) program of reduction but this time applied to the A-domain. Just as Topicalization and Relativization involve the common operation of A′-movement (despite being different in many other ways), Control and Reflexivization (and Passive and Raising) involve the common feature of A-movement (despite being different in many other ways). What distinguishes Control from Raising (and Passive) on this conception is not the primitive operations involved (they are identical in both cases) but the number of times A-movement (Copy and Merge) applies and the feature-checking positions through which elements are moved

[21] It is worth observing that Chomsky (1977) also tries to reanalyze deletion rules like Comparative Deletion in terms of Wh/A′-movement. In effect, Chomsky argues that deletion rules that show island like properties should be reduced to movement. This reduction serves to explain *why* such rules obey island conditions, the latter being a property of this operation by eliminating a redundancy in the theory of UG (see Chomsky 1977: 89).

(e.g. Control and Reflexivization transit through theta positions, unlike Raising and Passive). As in the case of Chomsky's (1977) thesis, this kind of reduction has explanatory virtues: why are PRO and reflexives c-commanded by their antecedents? Because they are tails of chains formed by movement and the head of a chain always c-commands its tail. Why must reflexives and (obligatory controlled) PROs be locally licensed by their antecedents? Because they are residues of A-movement and thus only exist if something (viz. the antecedent) has moved from there in the way typical of A-movement (e.g. obeying minimality and least effort).

Though reduction, if possible, is *always* methodologically favored because it enhances explanation, in the present context it has one additional benefit. If achievable it has the interesting consequence (interesting given considerations mooted in 1.2 above) of reducing the modularity characteristic of GB theories of UG. Binding, Control, Case checking and theta role assignment result from the same basic operations subject to the same conditions. What differs are the features checked. Thus, though grammars check many different kinds of features they do so using the same basic machinery, the operations Merge and Move subject to minimality. Thus, for example, case features and theta features are checked by merging (via A-movement) near case and theta assigning heads and Relativization, Topicalization, etc. by merging near Topic and Relative C^0 heads (via A′-movement). If this line of analysis is correct, then underlying the complexities of the many diverse linguistic relations sit two operations (viz. Merge and Move) and the conditions that they are subject to (e.g. minimality and (something like) subjacency).[22]

Given this line of thought, reduction has two alluring properties if successful: It increases explanatory power and it simplifies the structure of FL. As the latter is a precondition for addressing the evolutionary question of how FL might have arisen in such a relatively short time, it contributes to the program schematized in (1). However, though reduction is a required first step, it is still only a first step. The next step is to factor out those features of FL that are irreducibly linguistic from those operations and principles recruited by FL from other cognitive domains. This constitutes a second minimalist project.

Consider an example. Take the basic operation Merge. It is normally taken to operate as follows: It takes two constituents as input and combines them to form a novel constituent labeled by one of the inputs. Thus, a V can combine with a D to form an object labeled by the V: $\{\underline{V}, D\}$.[23] Merge is subject to

[22] If Move is actually an instance of Merge as proposed in Chomsky (2004), or the combination of Copy and Merge as proposed in (Chomsky 1995a), then we can reduce grammatical relations to various applications of Merge and feature checking.

[23] Underlining identifies the expression that names the output. Labeling amounts to identifying one of the two merging elements. It is not an operation that need "write" the name of one of the two expressions as a label. For our purposes, it is equivalent to $\{X, \{X, Y\}\}$ in current notation.

certain conditions. It is binary, it is subject to the Extension condition and its product has only one label. One can reasonably ask: whether this operation is "atomic"? Whether it is a primitive operation of FL or an instance of a more general cognitive operation? Why it merges at most two constituents and not more? Why it obeys the Extension condition? Why only one constituent labels the output? Why the merge involves labeling at all? What a constituent is? How it is different from Move? Etc.

All of these are reasonable questions, some of which have putative answers. For example, it is reasonable to suppose that an operation like Merge, one that "puts two elements together" (by joining them or concatenating them or comprehending them in a common set), is not an operation unique to FL. It is a general cognitive operation, which, when applied to linguistic objects, we dub "Merge." The Extension Condition, which applies to all structure-building operations in the grammar, is also plausibly a reflection of computational considerations that apply beyond the linguistic case. It has the property of preserving the structural properties of the inputs in the output. This is a "nice" property for a computational system to have because it avoids the revision of previously computed information (i.e. it makes structure building monotonic). Computations progressively add information. They never delete any. As grammars are computational systems (plausibly well-designed ones) we would expect them to be monotonic. Note that this reasoning explains why a computational operation like Merge obeys a condition like Extension. Extension is the linguistic expression of the *more general computational desideratum* of monotonicity and as such is not specific to FL.[24]

What of labeling? This is less obviously what we expect of computational operations. The labeling we see in FL leads to endocentric phrases (one's with heads). There is a lot of evidence to suggest that phrases in natural language are endocentric. Hence it is empirically reasonable to build this into the Merge operation that forms constituents by requiring that one of the inputs provide a label. However, there is little evidence that this kind of endocentric hierarchical structure is available outside FL. Nor is it obviously of computational benefit to have endocentric labeling for if it were we would expect to find it in other cognitive systems (which we don't). This suggests that endocentric labeling is a feature of Merge that is FL specific.[25]

We can keep on in this way until all the properties of Merge have been surveyed (we will do so in Chapter 5). However, the point here is not to analyze Merge's various properties but to illustrate what it could mean to distinguish first factor from third factor features. In the chapters that follow I will pursue

[24] This is essentially Chomsky's (2005a,b) No Tampering Condition. Extension is the requirement that Merge is only possible at the root. For further discussion see Chapter 2.

[25] There is some evidence to suggest that endocentricity facilitates language learning. See de Marcken (1996).

this strategy more single-mindedly. Recall that in the best possible case the truly distinctive features of FL are small in number (one or two) and the rest of its properties are actually reflections of language independent features of cognition. This is what we expect from a system that has only recently emerged.

Given (1), the project of finding the linguistically specific properties of FL is bounded on the input side by the operations and principles available to FL/UG that are not specific to language. It is bounded on the output side by the requirement that the (small number of) linguistically specific primitives together with the previously available mechanisms derive the generalizations of GB. This project thus gains teeth when considering the features of GB. If the project sketched in (1) is to be realized then many apparently language specific relations and operations will have to be exposed as special instances of third factor features. This is no small task given the many grammatical notions (critical to the GB version of UG and many minimalist accounts) that seem proprietary to language. Consider some examples.

In addition to Merge, which locally relates two expressions, Move is an operation that relates linguistic elements at a distance. A third operation is AGREE, which can relate linguistic expressions without "displacement" (e.g. agreement in existential constructions in English). Then there is binding, which allows two non-adjacent expressions to interact. Move, Bind and AGREE relations are ubiquitous in language but have no apparent analogues in other cognitive domains. In addition there is a plethora of relations like c-command, constituency, heads, maximal projections, etc., that also seem unique to language. These notions critically exploit the specific hierarchical structure characteristic of linguistic expressions and have no obvious analogues in other domains. Are these all primitives of FL or are they the outward manifestations in the domain of language of more general features of cognition? The logic of Darwin's Problem suggests the latter. The program is to show how this could be so.

One way of approaching this task is via questions like the following. What's the relation between Merge, Move and AGREE? There exist proposals that not all of these operations are primitive. Chomsky (2004) has proposed that Move is actually a species of Merge (ReMerge). An earlier proposal of Chomsky's is that Move is the composite of two other operations, Merge and Copy. As for AGREE, in GB non-proximate agreement was an indication of covert Move. More contemporary accounts eliminate covert operations and substitute (long distance) AGREE. Are either Copy or (long distance) AGREE language specific? If so, then they are part of the background operations that were exploited to form FL. If not, they are first factor primitives whose emergence needs explanation. Here are other relevant questions: Why does movement target constituents? Why does it obey Structure Preservation? Why are anaphors generally c-commanded by their antecedents? Why do moved elements generally c-command their launch sites? Why are sentences hierarchically structured? And so on. GB has provided us with a rich description of what sorts of properties FL has. The minimalist program aims to understand why it has these properties and not others. We answer these questions by showing how these facts about grammatical processes could have rapidly emerged from the combination of principles and operations not specific to language and one or two innovations (preferably one) specific for language. Borrowing from Chomsky (1965), we can say that GB is (roughly) descriptively adequate in that it (more or less) correctly describes the laws of FL. We can say that a minimalist hypothesis is explanatorily adequate if it explains how these laws could have emerged rapidly, i.e. by showing how a small addition specific to language combines with general cognitive principles to yield these laws.[26]

The two minimalist projects limned above clearly go hand in hand. Solving Darwin's Problem will require reducing the internal modularity of FL by showing how the effects of a modular system arise from the interaction of a common set of operations and principles. This then sets up the second question regarding the source of these operations and principles. It is hoped that most are expressions of operations and principles characteristic of cognition and computation more generally. The Minimalist bet is that these kinds of theoretical projects can be fruitfully pursued.

1.4 The structure of the book

The goal of this book is to develop a way of implementing these proposals. Much of my research since the mid 1990s has focused on developing a minimalist account of UG. I stress the indefinite article here. There are many analyses that fly under the minimalist flag and many different ways of understanding the goals of the program, often embodied in different technologies. Not surprisingly, the story that I develop is based on my earlier work and the theoretical and technical decisions embodied therein. One of these is of some moment in what follows: I assume that two central cases of binding, viz. local anaphora (Principle A) and obligatory control, are products of movement.[27] I have argued for these positions in other places.[28] This book presupposes that this sort of

[26] I would be inclined to go further and incorporate Embick and Poeppel's proposal that an explanatorily adequate account provide a solution for the granularity problem as well.

[27] For arguments in favor of these views, see Boeckx and Hornstein (2004, 2006), Boeckx, Hornstein and Nunes (in progress), Hornstein (2001, 2003, 2006), Kayne (2002), Lidz and Idsardi (1998), Polinsky and Potsdam (2002), and Zwart (2002) and references therein for some relevant literature.

[28] See Hornstein (2001, 2003, 2006) as well as Boeckx, Hornstein and Nunes (2007 and forthcoming).

movement approach is empirically and theoretically viable. It concentrates on showing how this assumption can serve to derive some of the basic properties characteristic of binding and control, most specifically the c-command requirement characteristic of the basic cases. Before grunting the details, here are the two other main ideas.

First, all grammatical relations are grammatically executed under Merge. For example, α theta marks β iff α merges with β. α controls β iff α merges with β. α locally binds β iff α merges with β. α case marks β iff α merges with β, and so on.[29] Clearly, for cases where, for example, antecedents are not sisters with the anaphors they bind, then the merging that establishes the grammatical relation must be followed by movement of the antecedent. This, in effect, adopts the idea going back to Kayne (1972) and Sportiche (1988) that doubling followed by movement is widespread in the grammar. The various locality conditions displayed within language are then reduced to conditions on movement (with movement itself being the product of Copy and Merge).[30] The bulk of the discussion concentrates on what the relevant conditions on movement are and how they are to be understood. As the larger aim is to address Darwin's Problem I try to show how these conditions on movement are just conditions on "nice" computations, thus understanding them as third factor properties.

Second, Merge is a species of concatenation and hierarchy in language is the result of combining concatenation with endocentric labeling understood in a Bare Phrase Structure way. Labeling so understood has the effect of closing the domain of lexical items under concatenation thus producing hierarchical structures. This closure has the effect of defining (syntactic) equivalence classes for a given lexical item (viz. all those that are labeled by the head). All items in this equivalence class are treated as the same by the computational system. I take endocentric labeling to be the principal "novelty" of UG, which, in combination with operations like Merge, Copy, Check Feature, and the various "nice" computational conditions these are subject to, yields many of the central properties characteristic of natural language.

These ideas and their consequences are developed in the following chapters. For quite a while, attempts to solve Plato's problem led to interesting conjectures about the structure of UG and deepened our understanding of FL. I believe that addressing Darwin's Problem can have a similarly stimulating effect. What follows is an attempt to make good on this hunch. As always it is for the reader to decide whether the attempt has been successful.

[29] Note that this is a necessary condition. Clearly all of these relations are asymmetric. Thus Merge must be as well. Labeling clearly introduces such asymmetry into the system and so will serve to distinguish, e.g. theta-marker from theta-markee even if part of the operation (Concatenate in Chapter 3) is symmetric.

[30] Treating Move as an instance of Merge (ReMerge) would serve equally well.

2 Deriving c-command

2.1 Introduction: The sources of c-command

Of the core syntactic relations in UG, none is more gregarious than c-command. It plays a key role in at least three different domains: binding, linearization and movement. Consider how.

All three principles of the binding theory exploit c-command in their definition of *binding*, binders being expressions that both ***c-command*** and are co-indexed with their dependents. More concretely, anaphors must be locally bound by their antecedents, pronouns cannot be locally bound by their antecedents, and R-expressions cannot be bound at all. In addition, pronouns interpreted as variables ("bound pronouns") are (typically) c-commanded by their antecedents.

Similarly, most (if not all) versions of the Linear Correspondence Axiom (LCA) are defined in terms of asymmetric c-command: thus α precedes β just in case α asymmetrically ***c-commands*** β.

Lastly, movement also crucially invokes c-command. For example, ECP-based accounts define antecedent government in terms of binding and the latter, as noted, is defined in terms of c-command. In addition, chains are defined in terms of c-command (links in a chain c-command one another) as is a central well-formedness condition on movement and/or chains, the minimality condition. Consider the latter, as it will be a focus of what follows.

Minimality restricts operations in the configurations in (1).

(1) Minimality: A movement operation cannot involve X^1 and X^3 over an X^2 which is identical to X^3:
$\ldots X^1 \ldots . X^2 \ldots . X^3 \ldots .$

A key feature of the above restriction is that it only applies when the relevant Xs are in a c-command configuration; in particular, X^2 blocks X^3 just in case it ***c-commands*** X^3.

This chapter aims to derive the fact that the c-command relation plays a role in all of these areas of grammar from (what I believe to be) more fundamental principles of grammatical organization. Given that "deriving c-command" has been a widely practiced sport of late, let me say a bit about how what I aim for here differs from other similarly ambitious projects (all of which, incidentally, I will shamelessly steal from to further the present enterprise). Epstein (1999) is the best-known (deservedly so) attempt to reduce c-command. However, in my opinion, though Epstein (1999) accomplishes a lot, it does not *deduce* c-command. Rather, it rationalizes it. Here's what I mean: it shows that c-command is a *natural* relation in the context of a Merge-based approach to grammar. Chomsky (2000) picks up on this theme and suggests another reason for why c-command is natural. Indeed, both these efforts postdate an earlier attempt to rationalize this relation in Chametzky (1996). What all of these approaches have in common is that their aim is not to show that c-command falls out from independent properties of UG but that c-command is a most *natural* relation given the way the grammar functions (i.e. given its basic operations, relations, and structures) and so we could expect it to be singled out for special regard.

Though I have nothing against this claim, I believe that it does not go far enough. And, in some sense, it goes a bit too far. Chomsky (2000) in particular does not recognize c-command as a mark of grammar, though it does take it to be a reflex of grammar. This emerges in the suggestion that Binding effects are products of interface operations rather than products of the grammar properly speaking. The reasoning goes as follows: That Binding is structured by c-command is no surprise as c-command is a *natural* relation of grammar and why shouldn't the interfaces exploit natural grammatical relations to do what they need to do.

This is too liberal for my tastes. C-command is a signature property of the kinds of hierarchically dependent relations that human grammars exploit.[1] As such, c-command sensitive "constructions" (e.g. Wh-movement, bound anaphora, etc.), I believe, reflect the most distinctive features of the human Faculty of Language. Consequently, for a process to be sensitive to c-command is sufficient reason (or very strong *prima facie* reason) for concluding that it is a product of the grammar. This is what I meant above when I said that c-command was a *mark* of grammar.

If one accepts this, the following is an obvious research question in light of the considerations of the previous chapter: What is it about the operations and principles of UG that lead to grammars that regularly exploit c-command when grammatical operations establish grammatical dependencies? Moreover,

[1] Others include head-based recursion typical of X′ structure and the locality conditions that constrain unbounded dependencies. The latter is discussed below and the former in the next chapter.

this question is also timely given recent minimalist inquiry. More specifically, some common assumptions currently in play (and which we review below) suggest that c-command is a necessary by-product of some kinds of Merge-based grammars. In effect, it is what falls out if UG is structured in a certain way. The aim here will be to specify what that way is.

A caveat before proceeding. The reader should be warned that what follows *presupposes* that the standard claims concerning the role of c-command in binding, linearization and minimality are essentially accurate.[2] The game played here takes these claims as fixed points and considers whether the role of c-command in various parts of the grammar can be derived from more basic assumptions. Put otherwise, what follows is an exercise in *theoretical* syntax.[3] That means that though it rests on an empirical base, its focus will be on the consequences of certain large bore theoretical assumptions that are prevalent in the literature; in particular, to show how they combine to give us an interesting conclusion about c-command. Work advocating the empirical utility of these assumptions is adverted to in the notes. I believe that one of the successes of the Minimalist Program is that these sorts of theoretical explorations are possible and (possibly) enlightening for it testifies to the deductive richness of its leading ideas.

The chapter is organized as follows. We discuss each of the major areas where c-command has proven to be grammatically central: Binding Theory, Linearization and Minimality. In each domain relations and operations are sensitive to c-command. For each case, I argue that the effects of c-command follow (or can be made to follow) independently and so c-command is not a primitive relation coded as such in FL/UG. The claim, then, is that c-command is a relation that one expects from a grammar organized in a particular fashion. It is a by-product of how FL/UG is structured rather than constitutive of its organization.

[2] "essentially accurate" does not mean "accurate in every detail." As will become clear in our discussion of binding below, it may well be that there are cases of binding without c-command. If so, this suggests that c-command is *not* a primitive, though it still remains to explain why it is such a very good *approximation*. This is discussed in more detail below.

[3] Theoretical syntax is distinct from formal syntax, though the two are often run together. I take the former to be concerned with the analytic connections between the leading ideas of a given program. When successful it leads to insight. The latter adverts to issues of technology, formal renditions of ideas often leading to clarification and explicitness. Though one hopes that insight goes hand in hand with explicitness, not every explicit proposal need be enlightening and not every enlightening proposal need be formal. At this given point of syntactic research, it is likely that any theoretical discussion will also be formal. Indeed, a good deal of theory will concentrate on examining the formal properties of our most effective accounts. This said, the distinction is worth keeping in mind as the aims of theoretical and formal work are not always the same.

2.2 Binding

2.2.1 *The standard case; single rooted sub-trees*

As noted, the Binding Theory (BT) highlights c-command in its binding requirements; all the relevant inter-nominal dependencies are among DPs related by c-command. Can the fact that interacting DPs line up in c-command configurations be derived? Perhaps.

Let's consider anaphors first. Principle A of the BT requires that an anaphor be locally c-commanded and co-indexed (bound) by its antecedent. Several authors have recently argued that the salient locality facts concerning anaphors follow quite naturally if we assume that the anaphor is a residue of overt (A-)movement.[4] Thus, for example, the acceptability of the sentences in (2) can be related to the acceptability of those in (3).

(2) a. John believes himself to be tall
 b. *John believes himself is tall
 c. *John would prefer for Mary to like himself

(3) a. John was believed t to be tall
 b. *John was believed t is tall
 c. *John would be preferred for it to be seen t (= John would be preferred to be seen)

In each case, the movement is illicit and so is the resulting binding relation. This would make sense if in fact the reflexive were a residue of overt movement.

This intimate connection between movement and anaphora is explicitly recognized in Chomsky (1981). Here the traces left by A-movement are categorized together with reflexives as anaphoric elements subject to principle A. Thus the parallel locality effects are simply reflexes of the common binding requirements to which A-traces and lexical anaphors are both subject. Note that this also explains why both NP-traces and anaphors are related to elements that *c-command* them.

The idea behind the more recent work retains the basic intuition in Chomsky (1981), but reverses the explanatory dependency; it is not that movement configurations are accounted for in terms of the binding conditions to which their outputs may be subject but that binding requirements of anaphors are explained in terms of movement operations that generate them. Why the switch? There are several good reasons for the change in perspective. Let's discuss them briefly in turn.

First the move to Minimalism involved an important change in theoretical perspective on traces. Within GB traces and other empty categories (like *pro*) are

[4] See, for example, Hornstein (2001), Lidz and Idsardi (1998) and Zwart (2002).

different in kind from their overt siblings. Traces are theory internal constructs and hence different in kind from regular lexical elements, including reflexives and pronouns. GB takes it as natural to understand UG as concerned with licensing these empty elements. Minimalism docs not share this conceptual motivation for it treats traces and lexical elements on a par. This is part of the central motivation for the Copy theory and reflects the minimalist commitment to the elimination of theory internal constructs. However, if there are no traces and no licensing condition specific to them then the GB strategy of linking the c-command conditions on movement to binding requirements on traces must be rethought.

Second, there is a technical problem. Chomsky (1981) argues that traces come in different flavors. In particular, A′-traces differ from A-traces in that the latter but not the former are subject to principle A of BT. Thus, only the latter are *required* by BT to have antecedents and so only these will be required to have *c-commanding* antecedents in overt structure. As a result, only A-movement will be forced to carry expressions to ascending c-commanding positions. In other words, the trace residues of A′-movement operations like Wh-movement are not anaphoric (they are R-expressions) and so the binding principles cannot explain why A′-movement *always* moves A′-expressions to positions that c-command their original launch sites in overt syntax (just like A-movements do).[5]

Third, the elimination of D-Structure (DS) and the reintroduction of generalized transformations (within the minimalist program) offers a new option for explaining why overt movement always targets c-commanding positions. Chomsky (1993) argues for the elimination of DS and for interspersing structure building and movement operations throughout the derivation. Cyclicity notions are incorporated via the requirement that phrase building always *extend* structure; the Extension Condition (EC).[6] This has the effect of restricting operations to the tops of phrases. (4) illustrates the Extension requirement.

[5] Note, I am not claiming that there was no account for this fact in earlier theory. It can be stipulated that variables must be c-commanded by their antecedents to be licit. However, this "explanation" has two drawbacks. First, it gives different accounts for why both A- and A′-movement target c-commanding positions and why these requirements parallel the requirements that anaphors impose on their antecedents. Second, the c-command requirement on variables holds at LF and so this fails to explain why A′-movement targets c-command positions in *overt* syntax. Third, that movement targets c-commanding positions appears to hold for every type of movement; Focus movement, Topicalization, VP fronting, Locative inversion, Scrambling, etc. Some of these, it has been argued, fail to form operator-variable structures at LF as they must reconstruct (see Chomsky 1995a, Heycock 1994, Saito 1989). These cases are not directly amenable to an account in terms of the requirements that operators and variables must meet.

[6] The Extension Condition was widely adopted in early minimalism. Other conceptions within AGREE/Attract-based theories have been explored which allow movement to target a non-root position. These theories stipulate that a probe must c-command a goal in order to probe (i.e. AGREE with) it. If movement requires AGREE (e.g. it is AGREE + EPP) then movement will target c-command positions (assuming that the agreement and EPP features are features of the

(4) a. $[_{\alpha} \ldots X \ldots] \quad [_{\beta} \ldots Y \ldots] \rightarrow [_{\beta}[_{\alpha} \ldots X \ldots][_{\beta} \ldots Y \ldots]]$
b. $[_{\alpha} \ldots X \ldots] \rightarrow [_{\alpha}X[_{\alpha} \ldots X \ldots]]$

(4a) merges two phrases to make a larger third phrase, while in (4b) a phrase is moved from within a previously constructed phrase and moved to the edge. In both cases, the resulting structure contains the previous phrases as proper subparts and so extends the structure. This would contrast with an operation like the one in (5) where the merger results in a "thicker" structure not an extended one as the resulting phrase fails to contain its inputs as subparts.

(5) a. $[_{\alpha} \ldots X \ldots][_{\beta} \ldots Y \ldots] \rightarrow [_{\beta}[_{\alpha} \ldots X \ldots] \ldots Y \ldots]$
b. $[_{\alpha} \ldots X \ldots] \rightarrow [_{\alpha}X \ldots Y \ldots]$

As Chomsky (1993) observes, EC is a very natural computational condition as it both restricts the grammatical "action" to the "tops" of the relevant computationally involved objects (thereby plausibly limiting "search") and enforces a monotonicity requirement on structure building operations.[7] At any given step, the relevant operations will leave the internal constituency of the inputs unchanged by only creating structure at the tops of the resulting phrase markers. At any rate, this conception of the cycle only becomes available once the idea that movement and phrase building operations can intersperse is adopted and this is only possible once the classical notion of DS is abandoned.[8]

For current purposes the example of interest is (4b) for it shows how Extension applies to movement operations.[9] Here X has moved from inside the αP and merged with it at the root. Only roots can be targets of operations subject to EC. As a consequence, if movements are subject to EC then a moved expression *must* c-command its launch site. Or to put this more tendentiously, the requirement that a moved expression c-command its launch site follows from the EC. Note moreover that this holds regardless of the kind of movement involved. More specifically, both overt A- and A′-movement will target c-commanding positions as both are subject to EC. In short, *if* structure building operations are subject to EC (a computationally attractive idea) then we *derive* the fact that movement results in structures in which moved expressions c-command

same head (i.e. the strong feature version of the EPP). In what follows we put these sorts of accounts to the side as they stipulate the c-command property that we aim to derive. See below for some further discussion of such accounts in the context of Tucking-In derivations.

7 Extension plausibly enforces a "no tampering condition" as well in that structure once built is never undone; once a constituent, always a constituent!

8 EC fits comfortably within a derivational conception of grammatical operations. It is not clear that it naturally comports with a representational conception of grammar. Thus, the discussion here, if persuasive, suggests that the grammar has at least some derivational aspects. This observation seconds a similar sentiment in Epstein (1999).

9 See Chapter 3 for a derivation of the Extension Condition from a more articulated understanding of the Merge operation.

their base positions, at least in standard cases of movement within single rooted sub-trees.[10]

Let's now drop the other shoe: if anaphors are related to their antecedents through movement, then we also derive the fact that they must be c-commanded by their antecedents![11] In short, if overt movement subject to EC mediates anaphoric dependencies, then it follows that anaphors (e.g. reflexives) must be c-commanded by their antecedents.

Let me put this in another slightly more abstract way. I proposed in Hornstein (2001) that the optimal way of grammatically coupling two expressions is via Merge, as this is the conceptually necessary grammatical operation.[12] If Move is just an instance of Merge (e.g. ReMerge) or the combined effect of Copy and Merge, then it is reasonable to suppose that inter-nominal dependencies are optimally coded by Move (as Move is just a species of Merge). It follows from this, plus the idea that (overt) movement obeys EC, that an anaphor (a residue of overt A-movement) will be c-commanded by its antecedent.[13]

What of the other Binding principles? There are two basic approaches each of which yield the same c-command restriction. Kayne (2002) proposes that pronominal binding, like anaphoric binding, is executed via overt movement. The basic idea is that bound pronouns are similar to doubled clitics. If so, the logic outlined above applies to bound pronouns and c-command is expected. Hornstein (2001, 2006) pursues another route to the same conclusion regarding c-command. It revives the original Lees–Klima theory but in a more contemporary minimalist setting. It is proposed that pronoun binding be parasitic on failed movement; licensed just in case movement is not. It involves a process of *pronominalization* in which a DP is replaced by a pronoun and *(re)merged* elsewhere in the tree. This process is how the grammar creates bound pronouns (i.e. pronouns interpreted as bound variables). Details aside, what is important for current purposes is that pronominalization is a structure building process in overt syntax. Hence, it is governed by EC. As such, it is expected that a bound pronoun will be c-commanded by its antecedent in single rooted configurations like (4b) above. Thus, it is expected that bound anaphors and bound pronouns will be c-commanded by their antecedents as (i) Merge (either as part of Move or Pronominalize) is involved in establishing the relation between antecedent and dependent in overt syntax and (ii) Merge is subject to EC.[14]

[10] We return to more complex cases involving sidewards movement anon.

[11] Once again, recall the caveat: this holds for the standard cases. Cases involving movement between sub-trees, so-called sidewards movement, are discussed below.

[12] Moreover, some grammatical relations are already discharged under Merge (e.g. theta relations, s-selection and c-selection). As some are so discharged, the optimal assumption is that *all* are so discharged.

[13] See Hornstein (2001) for a fuller discussion and the references in note 4.

[14] As is well known, there are empirical problems with the c-command requirement on bound pronouns. There exist cases of pronominal binding in which the antecedent does not c-command

This leaves Principle C. This is not the place to discuss Principle C in detail. Suffice it for now to observe that it is the reflex of the fact that interchanging the relevant pronoun and antecedent is a licit binding configuration. In short, Principle C takes effect where binding would have been licit. This is essentially the idea in Reinhart (1983) where it is proposed that the optimal way of co-valuing two expressions is via binding (viz. anaphoric binding or pronominalization where possible). This amounts to saying that α and β can be "accidentally" co-valued in a given structure γ just in case swapping α and β in γ is not a licit binding configuration (one that could have been formed via movement or pronominalization).[15] Tying the possibility of co-valuation on the *unavailability* of "binding" leads to "accidental" co-valuation being blocked just in case α and β are in a c-command configuration (recall that we have just reduced all cases of binding to operations in overt syntax, thus subject to Extension and therefore yielding c-command configurations). In short, if one adopts Reinhart's thesis (which relates principle C to binding), and the proposals above that Merge underlies both anaphoric and bound pronoun dependencies (via Move and/or Pronominalize respectively), then we explain why c-command conditions Principle C.

This takes care of the basic binding cases. As should be clear, it also accommodates the standard obligatory control configurations. In the canonical cases of obligatory control, controllers c-command the PRO they control. This immediately follows if OC PRO is an A-trace formed by movement.[16] If "PRO" is the residue of Move, then Extension is expected to play its usual role and enforce c-command between controller and controllee. Thus, like the anaphor cases discussed above, if OC PRO is the result of overt A-movement and the construction of (overt) syntactic structure is subject to EC then a controller must c-command the PRO that it obligatorily controls.[17]

It seems then, that the canonical cases of binding are expected to yield c-command configurations if their generation is tied to overt syntactic processes, movement being the default case. If anaphoric dependency is optimally executed via Merge/Move (and, perhaps, Pronominalize), and if these

its target. For review, see Hornstein (1995). This said, I will here ignore these cases, but see discussion below of Chinese reflexives.

15 Co-valuation can occur in two ways:

(i) x is assigned value v and x′ is assigned value v′ and v = v′
(ii) x is assigned value v and x′ is assigned the value assigned to x

(i) is "accidental" coreference and (ii) is binding. Reinhart's proposal is that (i) is possible just in case (ii) is not.

16 See Hornstein (2001), Boeckx and Hornstein (2004), Boeckx and Hornstein (2006) and Boeckx, Hornstein and Nunes (forthcoming) and references therein.

17 This also implies that non-obligatory control (NOC is not formed by movement), as it does not require that an antecedent of an NOC PRO c-commands it. See Hornstein (2001) and Boeckx *et al.* (forthcoming) for discussion.

structure-building operations are subject to extension, then c-command between an antecedent and its dependent must result.

2.2.2 *Sidewards movement*

Before moving on, let's consider some complications to this basic picture. To this point, we have considered the standard cases in which movement is between elements in a single rooted sub-tree. However, other possibilities exist given some current conceptions. For example, if there is sidewards movement (SWM) (for which I believe there is decent evidence), need binding be restricted to c-command in these cases? This is both an empirical and conceptual question and I discuss both aspects briefly here.[18]

Empirically, it seems that c-command does generally obtain in cases of SWM. For example, Nunes (1995) analyzes Parasitic Gap Constructions (PG) in terms of SWM. In such cases, an expression moves from an adjunct to a theta position and then to a CP position as schematically illustrated in (6).[19]

(6) $[_{CP}$ WH . . . $[_{TP}$ . . . $[_{vP}$ . . . $[_{vP}$ . . . t . . .$][_{adjunct}$ . . . t . . .$]]]]$

As the relevant adjuncts are typically thought to adjoin to *v*P or some other projection in the vicinity of *v*P, and as PGs involve some CP position as the ultimate landing site, we should expect the surface form of such constructions to end up in configurations in which the A′-expression c-commands all the positions through which it has moved. In short, if PGs are formed by overt sidewards A′-movement, then we expect the resulting structure to have the A′-element c-command its various previous launch sites. At least in cases like (6).

[18] It is often assumed that SWM must be *added* as a grammatical option and so grammars that allow it are more complex than those that do not. This, however, is incorrect. As Hornstein (2001) and Nunes (1995, 2004) observe, SWM is permissible unless specifically precluded. This exclusion can arise in a number of ways. We might stipulate that all grammatical operations must occur within single rooted sub-trees. Or we might only allow movement to occur subsequent to AGREE and the latter is restricted to elements c-commanded by the Probe (see Chapter 6 for discussion). However, short of these kinds of additions, a grammar that allows unconnected sub-trees in the course of a derivation and allows the simplest interpretation of Move (Copy and Merge/ReMerge) also allows an expression to be copied from one sub-tree and merged into another. This is SWM. So, the possibility of SWM follows from the least encumbered theory. Assuming its existence is the null hypothesis as preventing it requires additional stipulations.

[19] I abstract here from subject PGs, though the logic is the same as the real gap is always c-commanded by some expression in a higher A′-position. Hornstein (2001), Nunes (2004) and Nunes and Uriagereka (2000) argue (on both empirical and conceptual grounds) that movement is always from the adjunct to the matrix.

Another case of SWM discussed in the literature has a similar structure and conclusion. This involves sidewards A-movement resulting in adjunct control configurations as in (7).

(7) John saw Mary before/without/after PRO leaving the kitchen

If these involve SW movement, they yield configurations like (8).[20]

(8) $[_{TP}$ John T^0 $[_{vP}$ $[_{vP}$ John $[_{VP}$ saw Mary$]$ $[_{adjunct}$ without John leaving the kitchen$]]]]$

Here too, if we assume that the adjunct hangs below TP (the general assumption) then even with SWM the resulting structure will be one where the head of the chain c-commands the various positions through which it moved. It will c-command the adjunct "trace" (i.e. PRO) as the adjunct is c-commanded by the subject. The subject must c-command the adjunct if the adjunct is adjoined below TP and movement of the subject to Spec TP from Spec *v*P obeys Extension. It will also necessarily c-command the object "trace."[21] Thus, on the assumption that Merge must obey Extension and adjuncts merge below TP (e.g. *v*P being the likely target) then subjects will end up c-commanding their "trace" positions as they must raise to TP for case reasons. In sum, what derives c-command in these instances of SWM are three assumptions: that DPs must check case and/or WH features, that clauses have architectures in which theta domains are within case domains which are within A′-domains (i.e. the basic architecture of the phrase is [CP [TP [VP]]]) and that Merge (Move being a special case of Merge) is subject to EC.

There exist some interesting cases where SWM might not lead to c-command in overt syntax that are worth pointing out here. One involves sidewards A-movement to the right as in (9). (10) is an instance of (9).[22]

(9) $[_{TP}$ $[_{TP}$ PRO/reflexive$_1$ . . . .$]$ $[_{VP}$ V DP$_1]]$

(10) PRO/himself having to take a long shower made everyone late for class

In such cases, DP$_1$ moves sideways from the subject gerund to merge with the V. In particular, we get a derivation something like (11): First we build the gerund (11a). Then we select V, copy DP and merge it (11b). Then we merge the gerund and the VP. The gerund is the subject here, and the DP has merged

[20] That SWM is involved is proposed in Hornstein (2001, 2003).

[21] One point is worth mentioning: Hornstein (2001) and Nunes (2004) argue that adjunction obeys the EC (*pace* Chomsky 1993). Interestingly, the EC is critical in deriving CED effects in those cases where SWM from an adjunct is not licensed. That it also enforces c-command, is a side benefit of considerable interest, I believe.

[22] See Hornstein and Kiguchi (2003).

as complement of the V. Note that at every step Extension is adhered to (we are only adding to the tops of projections). However, because movement is sidewards, there is no requirement that we ascend the (connected) tree, and because the case position of the object is below that of the subject, there is no independent reason why the ultimate landing site of the object need be above that of the gerund (in contrast with the cases in (7) and (8) above).

(11) a. [DP . . .] V
b. [DP . . .] [V DP]
c. [[DP . . .] [V DP]]

Thus, here PRO/reflexive is not c-commanded by its antecedent, nor, apparently, need it be. If these cases are indeed cases of OC/binding, then it appears that there exist some licit cases of binding without c-command, which are derivable by a series of licit movement operations. Suffice it to say, that if these cases are derived as indicated, then c-command is *not* required for binding or control to be acceptable.[23] C-command is expected to obtain in standard cases of binding within single sub-trees (assuming that they are formed by overt movement), because the Extension enforces c-command in such cases. We also expect c-command to appear in those cases of SWM where the movement is to the left out of an adjunct adjoined below TP. However, cases like (9) (if they are indeed cases of SWM) suggest that c-command, though typically present, is not required for licit binding or control.

Another possible case where c-command does not hold involves cases in which the antecedent of a reflexive is within a DP. This sort of binding travels under the name of "sub-command" and occurs in many East Asian languages. Consider an illustration from Chinese.

An antecedent in Chinese can bind a local reflexive *taziji* even when contained within a DP (i.e. without c-commanding the reflexive).[24]

(12) Zhangsan de guiji hai-le taziji/??ta
Zhangsan de trick harm-perf himself/him
Zhangsan's tricks harmed himself/him

[23] The indicated derivation of cases like (10) is quite contentious (which does not mean to say that it is incorrect). Thus, one might resist the SWM derivation in favor of one in which the subject begins its derivational life as a complement and moves to the external argument position, much as in the case of psych-verbs. This too would allow for the noted inverse binding effects. Hornstein and Kiguchi (2003) argue that the derivation indicated in (11) holds for at least *some* cases, and this suffices for the point made here. However, if this does not prove to be correct, it does not affect the main point here as c-command would hold in the course of the derivation. This said, this case is of particular interest for *should* (9) require a derivation like (11), it would provide empirical support for the claim that c-command is not a primitive necessary structural condition on anaphoric binding.

[24] The long form of the reflexive *taziji* is a local reflexive and contrasts with the short form *ziji*. I'd like to thank Ming Xiang for the Chinese data.

(13) Zhangsan de shu zhi jiaoyu-le taziji/*ta
Zhangsan's book educated-PERF himself/him
Zhangsan's book educated himself/him

Note that here the reflexive is in complementary distribution with a bound/coreferential pronoun, as it should be if it is truly a locally bound reflexive. This sort of binding is easily derived assuming sidewards movement. The derivation is in (iii) (English glosses used):

(14) a. merge: [John self]
b. merge: [educate [John self]]
c. copy *John* and merge (sidewards movement): [John book]
d. merge: [[John book] [educate [John self]]]
e. Finish derivation in usual way to check case etc.
f. Delete non-case marked residues and add pronoun to reflexive morpheme:
[[John book] T [~~John book~~ [educate [~~John~~-self]]]]

With this derivation *John* becomes the antecedent for the reflexive though it does not c-command it. It is another illustration of the possibility of binding without c-command which is expected if reflexives are formed by movement and if sidewards movement is a grammatical option (cf. note 18).

The availability of sub-command in Chinese raises the question of what distinguishes English from Chinese. The key property that allows the Chinese facts and prevents analogous structures in English appears to be that Chinese reflexives require human antecedents, while English reflexives can be bound by non-human antecedents. This combines with the A-over-A principle to yield the difference in behavior.[25] This has the additional consequence of predicting that in Chinese sentences like (15) *Zhangsan* cannot antecede the reflexive, though it can antecede the pronoun. In other words, the effects in (12) and (13) are reversed.

(15) Zhangsan$_1$ de Mama guiji hai-le *taziji$_1$/ta$_1$
Zhangsan's mother harmed himself/him

To sum up: it seems that where c-command exists, its occurrence can be derived from more basic (computationally natural) assumptions concerning how phrases are constructed (i.e. the Extension Condition) and inter-nominal

[25] The proposal that sentences like *John's mother loves himself* is out because of something like the A-over-A principle is made in Kayne (1994: 25–26). There it is proposed that *John's mother* blocks *John* from being a possible antecedent as it is a more proximate potential antecedent. See Boeckx and Hornstein (2007) for implementation of this idea in a more general context. Chapter 3 discusses the A-over-A principle in more detail.

dependencies grammatically rendered (via Merge and Move).[26] The movement cases fall into two classes. Those that involve operations within a single rooted phrase marker rely on nothing more than Extension to derive the c-command requirement between dependent and antecedent. Those cases involving SWM, movement between sub-trees without a common root, assume Extension *plus* an assumption regarding the architecture of the clause (viz. that theta domains are contained within case domains which are contained within A′-domains) *plus* the assumption that movement is forced by the need to check uninterpretable features (viz. Greed (aka Least Effort)). Extension and Greed are core minimalist assumptions. That clauses are configured as indicated has been the reigning assumption since the earliest days of generative grammar (though *why* clauses must be so configured is unclear). Thus, from Extension (and a number of other conventional assumptions) we can derive that c-command will characterize the relation between Mover and launch site in the standard cases and, hence, antecedent and anaphor if the latter is just a special case of Move. Moreover, these same assumptions allow us to outline some cases where binding/control need *not* require c-command to be licit. Of course, if such cases exist, they provide an additional reason for treating c-command as derived from more basic features of UG. In sum, the present proposal provides a way of understanding both why binding and (obligatory) control require c-command in the canonical cases and why certain cases might be licit without c-command.[27]

2.3 Linearization

A second place where c-command has played a role is in the linearization operations of the grammar. Phrases are hierarchically organized objects. A standard method for linearizing phrases is via a version of the Linear Correspondence Axiom (LCA), which is an algorithm that imposes a left-to-right order on a

[26] Note it is critical that Move be understood as involving Merge as a subpart for this is what enforces Extension. Either the Copy Theory of Movement or the ReMerge approach to movement has the desirable effect. To the degree that c-command is understood to be derived in the manner outlined here, it constitutes further evidence in favor of treating Move as a derived, rather than as a basic, operation.

[27] There is another assumption that underlies the SWM derivations in which the mover ends up in a c-command position. The assumption is that theta, case and A′ domains are distinct. So, for example, an element receives its theta role within the lexical domain (effectively the V+*v*P shell), case within the TP and WH properties within the CP. The separation of domains plays a role in the above discussion for sidewards movement is licit from the adjunct sidewards to the theta domain of the matrix and then upwards for case, Wh-feature checking or both. For further discussion of how domains are organized within an MP grammar, see Grohmann (2003) and Chapter 7.

phrase's terminal elements.[28] A standard way of doing this involves an instance of c-command, asymmetric c-command (ACC).

(16) LCA: Linearize α before β if α ACCs β

(17) It is customary to understand α and β in (16) as ranging over terminals. Thus in a phrase marker like (18) we get the linearized order in (19).

(18) [John [likes her]]

(19) John>likes>her (where ">" means precedes)

There are well-known problems with (17). For example, how to linearize *likes* and *her* in (18) given that each c-commands the other. There are also various ways around this problem such as assuming that the correct input to (16) is a phrase marker where either *likes* or *her* or both have vacated the VP. If, for example, the LCA applies to (20) no similar "bottom of the tree" problem arises.[29]

(20) [$_{TP}$ John [$_{T'}$ [$_{vP}$ <John> [$_{v'}$ likes+*v* [$_{VP}$ <likes> her]]]]]

As is evident, this version of the LCA invokes c-command. However, it is by no means the only possible approach to linearization. Nor is it clear that c-command is *crucial* to the success of the algorithm. Here's what I mean. A linearized order is asymmetric in the sense that if α precedes β then β does not precede α. To induce a linearization from a hierarchical structure, one must find some asymmetric hierarchical relation among the elements that become linearized in terms of which the linearization can be executed so that the resulting linearization is asymmetric. ACC then is useful not because of its c-command part but because of the A(symmetric) part. The relevant question then is whether this is the *only* asymmetric relation that the grammar can pivot on to produce a linearization. If there are other asymmetric relations that the grammar has available beyond some version using c-command then they too can subserve linearization. With this in mind consider the following possibility. Assume that Merge is asymmetric (viz. not "α and β merge" but "α merges with β" or vice versa) and that (21) is the linearization algorithm.[30]

(21) LCA′: Linearize α before β if α has merged with β

LCA′ involves two departures from standard assumptions. First, that Merge is an asymmetric operation and second, that non-terminals are in the domain

[28] The LCA was first proposed in Kayne (1994). It was discussed in Chomsky (1995a) and Uriagereka (1999), among other places.

[29] Expressions in < > brackets are deleted and so not subject to the LCA. For a discussion of the LCA and the various problems that arise, see Hornstein, Nunes and Grohmann (2005).

[30] This idea is adopted from Epstein (1999) and is developed in Epstein *et al.* (1998) based on ideas of Kawashima and Kitahara (1995).

of the rule.[31] Before evaluating these departures from conventionality, let's consider how (21) applies to (20). We construct the sentence as in (22):

(22) a. Merge *her* with *likes*: [her likes]
b. Merge *v* with [likes her]: [*v* [her likes]]
c. Copy *likes* and merge with *v*: [likes+*v* [her likes]]
d. Merge *John* with *[likes+v [likes her]]*: [John [likes+*v* [her likes]]]
e. Merge T^0 with *[John [likes+v [likes her]]]*: [T^0 [John [likes+*v* [her likes]]]]
f. Copy *John* and Merge with TP: [John [T^0 [John [likes+*v* [her likes]]]]]

The order one gets doing this is illustrated in (23).

(23) a. her likes
b. *v*>[her>likes]
c. likes>[*v*>[her>likes]]
d. John>[likes>[*v*>[her>likes]]]
e. T^0>[John>[likes[>[*v*>[her>likes]]]]
f. [John>[T^0>[*John*>[likes>[*v*>[her>*likes*]]]]]

The italicized copies will delete and the left-right order of the lexical terminals will be *John>likes>her*.

Note that by tracking the history of the above asymmetric Merge operations, we end up with the correct linearization. Let's now turn to some details.

We assume that Merge is asymmetric. This is important for to induce a linearization (an asymmetric ordering), we need some asymmetric relation on which to piggy back. The LCA in (16) uses *asymmetric* c-command. But if we wish to remove c-command as a primitive of UG, then we need to find some other asymmetry in terms of which to leverage linearization, hence the assumption that Merge itself is asymmetric. Is this reasonable? I believe it is for the following reason. Assume that Merge, like all other grammatical operations, is last resort, i.e. that it only applies because it must. This means that when Merge takes place, some requirement of one of the participants is discharged via the merger. In the case of the merger of a V and its internal argument, it makes sense to think that a θ-feature of the V is being checked.[32] If we assume

[31] That Merge is asymmetric is perfectly reasonable for reasons noted below about Merge and Last Resort.

[32] Hornstein (2001) assumes that this is how thematic assignment occurs: a theta-feature of the predicate is transferred to the DP that it has merged to. Other options are possible. Thus Bowers (2005) assumes that it is a sub-categorization of V feature that is checked. What kind of feature is involved is of little moment for what follows. In fact *whether* features are involved is not particularly critical. What is important is that the operation asymmetrically affects the grammatical requirements of the participants. Whether the satisfaction of these requirements is tracked via "features" is of secondary importance. Thus, for example, one might argue (motivated by a GB sensibility) that merging a DP and a predicate satisfies some requirement

that α merges with β just in case α checks a feature of β, then in the case of internal arguments the object will precede the V.[33]

What of functional material: does *v* check a feature of V or vice versa? Does T check *v*P or vice versa? Here things are less clear intuitively. The verbal affixes of T *need* a verbal element to attach to, but just as clearly a stem *needs* verbal affixes to attach to them. It appears that a case can be made in either direction and this suggests that perhaps either is an option. Say that this is correct, then we encode a kind of "head" parameter and languages may differ on whether functional heads will appear linearized to the left or to the right.[34] If this is acceptable, then English is presumably a left headed language, e.g. *v* satisfies a requirement of VP, and T of *v*P, while the opposite is true in Japanese. If we assume this, then in English *v* merges with VP and so precedes it. In transitive clauses, the reason that the verb precedes the object is that V raises to *v* in overt syntax. As the lower copy is deleted, the verb will precede the object. Note too that the subject will precede the TP of which it is the Spec as it checks agreement (and maybe case) on the finite T when moving to its Spec. As the lower copy in *v*P deletes, the subject will be at the left edge of the clause.[35]

We need one more assumption to get the trains to run on time: assume that if α precedes β then α precedes *all* of β. For example, if *v* precedes VP then it

of the DP, not the V (i.e. it must acquire a theta role). For current purposes, it does not much matter which way the asymmetry breaks so long as Merge is asymmetric.

33 There is another way to think about this. In general, the asymmetry of Merge is reflected in which element projects the label. It has been generally assumed that if two elements merge, only one projects a label. As Chomsky (2000) notes, which element contributes the label is generally predictable. It is the element whose requirements the merger satisfies. If this is so, then one can take the merger to be the expression that does not project and the mergee to be the one that does. We can then rephrase the asymmetry as follows: α merges with β if β projects the label of {α, β}. Note, given this, for current purposes, we can assume either that Merge itself is asymmetric or Merge together with whatever is responsible for labeling induces asymmetry (thanks to Paul Pietroski p.c. for this point). Chapter 3 focuses on the central role of Labeling within FL. If the line of reasoning outlined there is correct, then linking linearization with labeling would be very natural. Consider the following reasoning: Labeling introduces asymmetry into the grammar. Linearization requires asymmetry to be operative. It is natural to hope that the asymmetry provided as part of the basic architecture of FL via Labeling is also exploited by the interface systems to linearize phrase markers. This way of understanding matters has one curious consequence given conventional assumptions: it will typically be the case that VP merges with *v* and so should precede it. If all the DPs vacate the lexical shell, however, this need not result in mandatory SOV order. See below for further discussion.

34 The utility of re-introducing a head parameter is argued for in Saito and Fukui (1998). The proposal here restricts the "parameter" to functional heads and so leaves Specs to the left of all heads as they check features on the heads they are specifiers to.

35 If T checks case on subject DP then why do subjects sit at the left edge of TPs? Why, in other words, are they the mergers and not the mergees? A possible answer: even if T checks a feature on DP, DP checks *more* features on T: both case and agreement. Note, possibly, there is no case checking at all (cf. Chomsky 2000). If so, only the Agr features of T are checked (with case being a reflex of this operation) and no problem arises.

precedes all the elements of VP. This too seems natural, though other options are conceivable.[36]

Taken together, these procedures will result in linear order tracking the history of Merge and Move (which is just Merge again) operations (as in Epstein *et al.* 1998). Importantly, if this way of coding linear order is viable, then we can treat asymmetric c-command as an emergent, rather than a fundamental feature of linearization. The linearized order of expressions reflects their history of merger. The crucial assumption underlying this approach to linearization is the assumption that Merge itself is asymmetric. This premise removes the need to leverage the asymmetric linearization relation via asymmetric c-command. Or, to put this another way: to generate an asymmetric ordering we need an asymmetric pivot. If we assume that Merge is symmetrical, then asymmetric c-command provides the necessary fulcrum. Removing c-command requires providing an alternative asymmetric relation on the basis of which linearization can be defined. An asymmetric conception of Merge provides the requisite relation and c-command in the guise of *asymmetric* c-command can be dispensed with for purposes of linearization.

Two points are worthy of note before moving on.[37] First, if we assume that linearizations must be total for derivations to converge, then operations like Tucking-in must be prevented.[38] To see this, consider the derivations illustrated in (24)–(25).

(24) $[_{X2}$ YP $[_{X1}$ X^0 [. . . .ZP . . .]]]

(25) a. $[_{X3}$ YP $[_{X2}$ ZP $[_{X1}$ X^0 [. . . .ZP . . .]]]]

b. $[_{X3}$ ZP $[_{X2}$ YP $[_{X1}$ X^0 [. . . .ZP . . .]]]]

(25a) is derived from (24) by Tucking-in. Assume that it moves (copy+merges or remerges) and checks a feature of X^0. If linearization tracks Merge, then in (25a), ZP will precede X1 as this is what it has merged with. YP will also precede X1 as it too has merged with X1 in (24). The problem is that YP and ZP in pre-X^0 positions are unordered as neither has merged with a constituent containing the other. Compare this Tucking-in derivation with the one in (25b). Here, ZP merges to X2 and so precedes it. Thus, it is ordered with respect to YP as the latter is contained within X2 (recall, if α merge with β then all of α is linearized before all of β).

[36] Chapter 3 proposes that Merge is actually a species of concatenation defined over atoms. If this is so and if linearization piggy backs on Merge, then linearization is essentially concatenate-left (and the relevant precedence notion is not "precede" but "immediately precede"). Interestingly, if Merge just is concatenation then the fact that if α precedes β then α precedes *all* of β follows without stipulation as atoms cannot inter-collate.

[37] The relation between linearization and Extension noted here was discussed in Kawashima and Kitahara (1995) and incorporated in Epstein *et al.* (1998).

[38] See Richards (2001) for discussion of Tucking-in. The assumption that linearizations must be total is quite standard, cf. Kayne (1994) and Nunes (1995).

The second thing to note is that the derivation of (25b) from (24) obeys Extension while the one in (25a) does not. We noted in the earlier section on binding that the Extension Condition has the nice property of enforcing c-command on overt movement. It plays an equally valuable role here in assuring total linearizations; so long as derivations adhere to Extension, linearization can track the history of Merge (asymmetrically construed). The reverse is also true: if linearization must piggy-back on the *history* of Merge, then Merge must obey Extension. This is worth emphasizing. In a grammar with Tucking-in we can define a linearization if we avail ourselves of c-command. For example, in (25a), YP asymmetrically c-commands ZP so if linearization were stated in terms of asymmetric c-command a grammar with Tucking-in could provide a total linearization. If, however, c-command is a derived notion, as it is here, then it seems that we need to assume the Extension Condition.

Though many details remain to be worked out, I will assume that the general strategy limned here underlies the linearization processes. The general conception that Merge is asymmetric fits well with the idea that grammatical operations are last resort and that labels can be predicted (see note 33). Interestingly, understanding Merge to be asymmetric allows for treating linearization as parasitic on the (successive) Merge operations themselves rather than the asymmetric c-command configurations that result. This is essentially Epstein *et al.*'s (1998) observation, which we adopt here. If correct, then the fact that linearization tracks asymmetric c-command is an emergent property of how phrases are constructed in a grammar whose operations respect last resort (viz. Merge is asymmetric), contain labels and respect Extension (monotonicity).[39]

[39] There are other conceivable ways of attaining the same result. Jairo Nunes (p.c.) suggests the following linearization algorithm to replace the LCA:

(i) The Default Linearization Algorithm (DLA): If α triggers merger with β then α must precede β if a total order obtains.

DLA is understood as follows: when a head H merges with its complement XP (and H projects) (i) gives the order H>XP. If YP now merges as specifier of H, i.e. H triggers merger of YP, then H should precede YP. However, if H precedes YP there is no total linearization as the order of YP and XP is unspecified. So (i) fails to obtain and the second option, YP>H is taken. Then by transitivity, YP is ordered with respect to XP and the order YP>H>XP is obtained. This is reminiscent of the argument above against Tucking-in: only if the specifier precedes the head can a total linearization be obtained.

The logic underlying the DLA gains some support from the view of Merge developed in Chapter 3. It is there proposed that only heads actually merge, so if H has merged with XP and then YP merges with the result, then YP only actually merges with H. This fits well with the logic of the DLA above. Here is not the place to argue for a specific version of the linearization algorithm. Suffice it to say that the DLA also dispenses with ACC and so serves to remove the need for c-command as part of the linearization process and so if correct, it suffices for current concerns.

2.4 Minimality and c-command

Consider now a last area of the grammar that relies on CC. Consider (1) repeated here as (26).

(26) Minimality: A movement operation cannot involve X^1 and X^3 over an X^2 which is identical to X^3:
$\ldots X^1 \ldots X^2 \ldots X^3 \ldots$

This condition only holds if X^2 c-commands X^3. This restriction is illustrated in examples like the Superiority cases in (27) and (28).[40]

(27) a. John wondered who books about what impressed
b. John wondered what whose mother said

(28) a. *John wondered who what impressed
b. *John wondered what who said

(28a,b) are standard Superiority effects. They can be analyzed as violations of minimality as the object *who* moves over the subject *what* on its route to CP. (27a,b) are not similarly unacceptable as they do not violate minimality *on the assumption that the latter only holds between c-commanding elements*. In (27a) *what* is buried within a DP and so does not CC *who* and in (27b) *whose* is inside the subject DP and so does not CC the launch site of *what*. As such neither blocks the movement of the object WH to CP.

Other examples make the same point. Consider some cases of A-movement. English allows raising over an intervening experiencer. Icelandic forbids this.

(29) John seems to Mary to be tall
(30) *Hestarnir virdast mer vera seinir
the-horses seem me-Dative to-be slow

The difference can be attributed to the fact that in Icelandic the experiencer carries dative case while in English it is object to the preposition *to*. In English, therefore, at the point where *John* raises to Spec T, *Mary* does not c-command it as it is buried within the PP.[41] In Icelandic, in contrast, the experiencer carries

[40] These sorts of cases were noted in Bošković (1999) where an Attract Closest analysis along the lines outlined here is developed. Richards (2001) offers different judgments. For what it is worth, I find (27b) better than (27a) and I find both better than the sentences in (28). In what follows, I assume that Bošković's characterization of the data is correct as it helps to illustrate the logic of minimality. One last point: there exist analyses of Superiority effects that do not rely on minimality, some of which I am quite partial to. For discussion, see Hornstein (1995) chapter 7 and references therein.

[41] This is proposed in Kitahara (1997).

dative case and is not within a PP.[42] Consequently at the point of the derivation where *John* wants to raise to Spec T, the dative experiencer intervenes blocking the ascent. The relevant structures are indicated in (31a,b).

(31) a. [T^0 [seems [$_{PP}$ to [Mary]] [Bill to be tall]]]
b. [T^0 [seems Mary [Bill to be tall]]]

A similar logic applies to control structures such as those in (31c,d).[43]

(31) c. *John persuaded Mary PRO (= John) to wash himself
d. John vowed to Mary PRO (= John) to wash himself

In (31c) *Mary* intervenes blocking the movement of *John* to Spec T. *Mary* does not block this movement in (31d) for *Mary* resides within a PP and so fails to c-command *John* at the point where movement to Spec T applies.

Enough illustration. Let's assume that the descriptive generalization in (26) is correct and see if we can account for *why* minimality only holds between expressions in c-command configurations. To start, let's consider why minimality holds *at all*. Why *should* dependencies be subject to this sort of restriction? The intuition behind Rizzi's original proposal is that grammars prefer shorter dependencies to longer ones. In other words, what minimality codes is a preference for relations among elements/positions to be as *short* as possible.[44] The next question then is: how do grammars evaluate distance? How

[42] This same approach might be extendable to the failure of raising over experiencers in Romance noted in Chomsky (1995a). Here the experiencer is marked with *á* which can function as either a case marker or a preposition. If when marking an experiencer it is functioning as a case marker, then it should block raising across it, as in Icelandic, as Chomsky reports. However, if it functions as a preposition, then it should pattern like English. It seems that this dual option is in fact realized and that speakers differ as to whether they can raise over an experiencer marked with *á*. Interestingly, the differences go away once the experiencer is cliticized. So whereas (i) garners conflicting judgments, cases like (ii) are judged uniformly acceptable (in fact, this is also true in Icelandic where experiencers, when coming in pronoun form, cliticize as well).

(i) Jean semble a Marie être intelligent
(ii) Jean me/le(?) semble être intelligent

The uniform acceptability of cases like (ii) makes sense if what cliticization does is merge the pronoun with the head with which it cliticizes thereby removing it as a c-commanding intervener.

[43] This adopts the movement theory of control, as in Section 2.2. For discussion see Hornstein (2001) and Boeckx *et al.* (in progress) and the references therein.

[44] See Rizzi (1990). Chomsky and Lasnik (1993: 89–90) describe the "basic and appealing intuition that lies behind the principle of Relativized Minimality" as follows:

> The basic intuition is that the operation Move α should always try to construct the "shortest link." If some legitimate target of movement is already occupied, the cost is deviance. We may regard this as part of the general principle of economy of derivation.

This is one of those conditions that have a natural computational rationale in that it circumscribes dependency relations. As (unbounded) dependencies can be computationally quite challenging, limiting their range makes good computational sense. For some discussion of the computational costs of grammatical dependencies see Berwick and Weinberg (1984).

do grammars compute the length of a dependency so that shorter ones trump longer ones? How, in other words, is distance grammatically computed?

A natural reply is: grammars measure distance by the nodes intervening between the related expressions. This is called a *path*.[45] So, for example, in (32a) the path of the *what* targeting C^0 is the set of nodes {VP, *v*P, TP, CP}, as these are the maximal projections that dominate the launch site of *what* (in VP) and its landing site (in CP).

(32a) $[_{CP}$ what C^0 $[_{TP}$ John T^0 $[_{vP}$ *v* $[_{VP}$ buy what]]]]

Paths provide a measure of the distance between two expressions in a phrase marker.[46] The preference for short dependencies can be recast as the maxim that path length should be minimized. So, for example, if some expression moves to check a feature of the target, say a Wh feature on C^0 in (32a), then the grammar wants to accomplish this with the shortest possible move.

That grammars choose shortest solutions to grammatical requirements is by now a theoretical dogma. Let's treat it with the respect that it deserves and assume it to be true. The next question is: how do grammars *compare* paths? I ask this because the obvious answer is almost surely the wrong one. This reply says that paths are measured by the elements that they contain and that the shortest one contains fewer elements. For example, the path from *who* to C^0 is shorter than the one from *what* to C^0 in (32b).

(32b) $[C^0$ $[_{TP}$ Who_1 T^0 $[_{vP}$ t_1 $[_{VP}$ buy what]]]]

The path of the former is {TP, CP} while that of the latter is {VP, *v*P, TP, CP}. Thus, one might say, the former has measure 2 while the latter has measure 4. As $2 < 4$, the first path is shorter than the second.

Though perfectly reasonable, this approach is almost certainly *incorrect*. The reason is that one of the more basic features of grammars is that they *do not count*. It seems that grammars don't have the wherewithal in general to distinguish number of operations, elements, etc. This is what lies behind the absence of mirror image rules, for example (grammars cannot express rules that say take a string numbered "1,2,3,4,5" and turn it into the string "5,4,3,2,1.") or the fact that whereas *ad*-jacency or *sub*-jacency is a regular relation, 3-jacency is not (affect the *next* thing is ok, but not affect the *third* thing). In effect,

[45] See Kayne (1984), May (1985) and Pesetsky (1982).

[46] The reader should observe that paths are being used here simply as units of measurement. There is no "path principle" or "path condition" being proposed. Rather paths provide a natural way of specifying a notion that is generally assumed but not generally defined: that a dependency has a measurable span. This span is measured in path size, rather than in parsecs, light years or meters. In effect, then, paths are units of phrasal distance and these are the units minimized by the Shortest Move/Minimal length condition injunction.

the absence of counting properties in grammars lies behind the ubiquitous observation that grammatical processes are structurally dependent rather than linearly dependent operations. If this is correct (a very safe bet!), then it implies that the answer above must be wrong for it measures path length by counting and this, we have seen, is something that grammars don't do.

So, if grammars cannot count but they must nonetheless measure path lengths (lengths of movements), how do grammars do this? One way is to use Boolean measures: the relative size of two sets is fixed if one is a proper subset of the other.[47] Thus, in the example above, the *who*-path ({CP, TP}) is a proper subset of the *what*-path ({CP, TP, *v*P, VP}) and so it must be shorter. Note, no counting here. If we assume that grammars compare path lengths by computing the subset relations among the various paths, then we can *deduce* that minimality is constrained by c-command. Let's see how.

The relation between c-command and path length becomes clear if we consider one more case and compare it with the one just discussed. Say that we had a structure like (32c).

(32c) $[_{CP}\ C^0\ [_{TP}\ [_{DP} \ldots Wh2]\ T^0\ [_{vP}\ t_1\ [_{VP}\ V\ Wh1]]]]$

Now compute the paths of Wh1 and Wh2 to C^0. P(Wh1) = {VP, *v*P, TP, CP}. P(Wh2) = {DP, TP, CP}. Observe that neither is a proper subset of the other. Thus P(Wh1) contains *v*P and VP (which are not elements of P(Wh2)) and P(Wh2) contains DP (which is not an element of P(Wh1)). Thus, though the measure of P(Wh2) is less than that of P(Wh1) neither is longer than the other if we compare paths in a Boolean fashion.

Consider one more case. Contrast P(DP1) and P(DP2) which target T^0 in (32di) and (32dii).

(32d) i. $[_{TP2}\ T^0\ [_{VP}$ seem $[_{PP}$ P DP2] $[_{TP1}$ DP1 . . .
ii. $[_{TP2}\ T^0\ [_{VP}$ seem DP2 $[_{TP1}$ DP1 . . .

In (32di): P(DP1) = {TP1, VP, TP2}, P(DP2) = {PP, VP, TP2}. Neither is a subset of the other and so neither path is shorter than the other. In (32dii): P(DP1) = {TP1, VP, TP2}, P(DP2) = {VP, TP2}. Clearly, P(DP2) is a subset of P(DP1) and so it is shorter.

[47] This holds for finite sets. Paul Pietroski (p.c.) observes that the restriction is actually stronger than non-counting. Two infinite sets can be the same size even if one is a proper subset of the other. The two sets may nonetheless be equinumerous as their members can be put in one-to-one correspondence. To derive the results below, we must assume that grammars not only do not count, but that they are restricted to Boolean assessments of grammatical options. See Chomsky (1957) for some discussion of this.

These two cases represent the raising and control examples discussed above where DPs within PPs fail to block movement.

At risk of stating the obvious, let's note what this shows: *if* grammars prize shorter dependencies over longer ones and *if* UG uses Boolean resources to evaluate grammatical options, then the way length is computed must be in terms of subset relations. The above proposes that what is so measured are paths, the set of maximal projections that dominate the launch site and the target. Grammars prefer those moves with the shortest "Boolean" paths. To be so comparable, the paths being compared must involve elements that c-command one another for failure to c-command results in paths that are not in subset relations and so are neither longer nor shorter than each other using a Boolean measure. Or, to put it more tendentiously: we have just derived the fact noted in (22) that c-command conditions minimality. Moreover, we have arguably explained *why* c-command should matter. Minimality is *not* itself the basic notion. Shortest dependency is (see note 44). Minimality conditioned by c-command is what shortness entails in a grammar restricted to Boolean measures. This, however, only becomes evident once we try to understand how grammars compute distance. Once we specify that the "unit of distance" is the path, it becomes clear why minimality should play a role in enforcing shortness and why, with respect to minimality, only c-commanding elements should be relevant.[48]

2.5 More on paths

We have outlined how paths can be used to measure distance. It is time to make this discussion a bit more precise, elaborate some further consequences, and consider some alternative ways of building them. The discussion above assumes a conception like the one in (33).[49]

[48] A confession: after many hours of thinking about it, I could not come up with any way of computing distance between two arbitrary points in a phrase marker (or any graph) that did not reduce to something like a path. I am tempted to say that the only way to measure distance in a hierarchically organized network is in path like terms (i.e. measuring distance in terms of nodes separating the relevant points). This said, there are many superficially different kinds of paths depending on what one includes: all projections, only maxPs, only functional projections, etc. We discuss this further immediately below and in Chapter 3.

[49] Other definitions are possible and would work equally well. For example, one could define a path of movement as in (i):

(i) Path: The target in union with the set of nodes dominating the mover.

In an example like (32di) the path of DP1 and DP2 would be the same as in the text under this definition. The main difference between (i) and (33) is that if we assume that domination is not reflexive, then it is conceivable that in some cases, there is no maxP dominating the target (see the discussion of sidewards movement in Section 2.5.5. below). If so, the path given the definition in (33) would not include an entry for the target while the one in (i) would. We return to this below.

(33) Path: a path is the set of maximal projections (XPs) that dominate the target or the launch site.[50]

This conception has some potentially interesting ramifications. Let's consider them briefly.

2.5.1 *The A-over-A condition*

Given (33), the A-over-A (A/A) condition reduces to minimality.[51] To see this, consider a typical A/A configuration.

(34) $[_{\text{TargetP}} \ldots \text{Target}_{\text{B-feature}} \ldots [_{\text{BP1}} \ldots \text{B1}^0 \ldots \text{BP2} \ldots] \ldots]$

In (34), BPs carry some B-features that need checking against the B-features of the target (or, if you prefer, some B-features of the target need to be checked by B-features of the BPs). Now consider the paths of BP1 and BP2. P(BP1) = {TargetP} while P(BP2) = {TargetP, BP1}. Thus, P(BP2) is a superset of P(BP1) and thus the latter is shorter than the former. Thus, by minimality, movement of BP2 out of BP1 should be barred.

The A/A principle is one of the more venerable within generative grammar (Chomsky 1964). Its effects can be seen in cases like multiple scrambling in Japanese. In Japanese it is possible to scramble a clause or a phrase. Moreover, multiple scrambling is possible. Given this, what happens if one tries to scramble both a clause and a phrase it contains? Is this possible? Yes, but only if one scrambles the clause first and then scrambles the phrase. The reverse order is prohibited.[52]

[50] We assume that domination is non-reflexive: a node does not dominate itself. Technically the definition of (33) is as follows: The path of α is the union of the set of nodes that dominate the target of α and the set of nodes that dominate the launch site of α.

[51] That minimality should extend to the A/A condition is noted (and used) in Boeckx (2003a: 66ff). A position similar to the one developed here is outlined in Fukui (1997). Željko Bošković (p.c.) notes that there is a tension between the A/A principle and the requirement to carry as little material as possible under movement. (Let's dub this "The Lightest Load Principle" (LLP). It is suggested in Chomsky (1995a), though never formalized.) This tension is partially resolved in the next chapter where a version of Chomsky's pied-piping intuition is developed. However, it is not clear what the significance of the tension is. The evidence for LLP comes largely from covert movement operations where pied-piping restrictions are relaxed. However, current single cycle theories eschew covert movement. For overt movement, the status of the LLP is empirically troublesome as it would preclude the kinds of pied-piping attested in Natural Language (e.g. moving a PP containing a Wh or moving a whole DP in languages where left branches can extract).

[52] The example in (35a) is simplified in that it is not clear that the scrambled object *Hanako-o* has scrambled out of the scrambled clause. However, that this is possible is attested by sentences like (i) where the subject intervenes between the scrambled clause and the scrambled object.

(i) $[_{\text{obj}}$ Hanako-o] John-ga $[_{\text{CP}}$ Taro-ga t_{obj} nagutta to] Mary-ni t_{CP} itta
Hanako-acc John-nom Taro-nom hit C^0 Mary-dat said

(35) a. $[_{obj}$ Hanako-o] $[_{CP}$ Taro-ga t_{obj} nagutta to] John-ga Mary-ni t_{CP} itta
Hanako-acc Taro-nom hit C^0 John-nom Mary-dat said

b. * $[_{CP}$ Taro-ga t_{obj} nagutta to] $[_{obj}$ Hanako-o] John-ga Mary-ni t_{CP} itta
Taro-nom hit C^0 Hanako-acc John-nom Mary-dat said
"That Taro hit Hanako, John said to Mary"

Why is (35b) unacceptable? Because it violates the A/A principle.[53] Hitherto, minimality and A/A have been considered to be separate conditions on movement. One side benefit of the current analysis is that the A/A reduces to an instance of minimality, more specifically to the general idea that grammars prefer shorter dependencies (i.e. that grammars minimize path length).

Consider one more example of the A/A from English that illustrates the same point. Consider a case of multiple Wh movement such as (36):[54]

(36) a. Which $person_1$ did you ask me [[how many pictures of $t_1]_2$ Bill took t_2]

b. [How many pictures of $t_1]_2$ did you ask me [which $person]_1$ [Bill took t_2 (= (how man pictures of t_1))]

(36a) is somewhat marginal. However, whatever its status it clearly trumps (36b) which is an incomprehensible lexical jambalya. We can account for the contrast between the two by noting that the latter violates the A/A while the former does not. Note that to derive (36a) the container Wh is moved first

It is also possible to scramble an embedded object over a subject oriented adverb that marks the left edge of a clause after the clause has been scrambled. The reverse is not possible. The contrast is exhibited in (ii) and (iii).

(ii) $[_{obj}$ Hanako-ni] orokanimo takarakuzi-ga t_{obj} atatta-to John-ga t_{CP} iihurasiteiru
Hanako-dat stupidly lottery-nom hit C^0 John-nom rumored

(iii) *[takarakuzi-ga t_{obj} atatta-to] Hanako-ni orokanimo John-ga t_{CP} iihurasiteiru
Lottery-nom hit C^0 Hanako-dat stupidly John-nom rumored
Lit. Stupidly, John rumored that the lottery hit Hanako
"Stupidly, John rumored that Hanako won the lottery"

Thanks to Masaya Yoshida for the brief tutorial on Japanese scrambling and to Tomo Fujii for help with relevant examples. The original examples are based on Kuno (2004) who analyzes these examples in terms of a generalization on outputs proposed in Müller (1996).

53 Kitahara (1997) also provides an A/A account of the facts in (35).

54 Similar cases are discussed in Fukui (1997) where it is observed that the contrast between (36a,b) is the main empirical support for the Proper Binding Condition (PBC). As Fukui (1997) notes, the PBC has two problems: (i) it fits poorly with minimalist assumptions (where the Copy Theory of Movement eliminates (or, at least, severely blurs) the distinction between traces and regular lexical items) and (ii) it seems to be empirically inadequate. Sentences like *How proud of Bill is John* violate it if we assume the subject internal predicate hypothesis. As Fukui (1997) further notes, and we repeat here, these cases can be adequately handled as A/A violations.

and then the contained Wh is fronted. To derive (36b) one first moves the contained Wh and only then moves the container WH. This violates the A/A (and minimality). I leave it as an exercise to the reader to develop the details and see that this is correct.[55,56]

2.5.2 *Minimal domains as exceptions to minimality*

Consider now a second consequence. It has been a staple of recent grammatical theory that minimality only applies to expressions in different domains. Or, to put this positively, expressions in the same domains do not interfere with one another, do not impose minimality restrictions on one another's movements. Why this should be so, however, has been theoretically disconcerting. Let's see why.

In Chomsky (1995a) it was assumed that movers in the same minimal domain (MD) are equidistant from any targets and that targets in the same MD are

[55] Interestingly, one more assumption is required. We need to assume, as we did also for the Japanese case, that once the un-interpretable features of an expression are checked they no longer "count" for minimality. Thus, the fact that the container has checked its relevant feature (scrambling or Wh) allows a contained expression with that same unchecked feature to move. This suggests that Chomsky's (1993, 1995a) analysis of weak Wh islands in terms of minimality is incorrect as they involve examples with a Wh in medial CP whose features have already been checked. Similar remarks extend to Fukui's (1997) treatment of Wh Islands as A/A violations. In addition, this suggests that Saito and Fukui (1998) is incorrect in taking scrambling to be non-feature driven movement, for if this were so it is unclear why scrambling the container then frees the contained.

[56] A similar analysis extends to cases of A-movement. Consider an example of "possessor raising" in Japanese. Japanese allows multiple accusatives in cases where there is a kind of inalienable possession.

(i)	a.	Hirohisa-ga	Masaru-no	atama-o	tata-i-ta
		Hirohisa-nom	Masaru-gen	head-acc	beat-particle-past
	b.	Hirohisa-ga	Masaru-o	atama-o	tata-i-ta
		Hirohisa-nom	Masaru-acc	head-acc	beat-particle-past
		"Hirohisa beat Masaru's head/beat Masaru on the head"			

(ib) shows two accusatives, as a result of possessor raising of the genitive from within the complex DP in (ia). Now consider a case of A/A. What happens to (ib) if we passivize?

(ii)	a.	Masaru-ga	atama-o	tatak-are-ta
	b.	*Atama-ga	Masaru-o	tatak-are-ta
		head-nom	Masaru-acc	hit-passive-past
		"Masaru was beaten on his head"		

(iia) is fine as the contained element is raised after the accusative on the containing DP is checked. Then, the case movement of the contained DP is fine. However, the converse is not. In (iib) the contained DP *Masaru* is case checked and then the container is passivized. This violates the A-over-A condition. Note, as above, we assume that once case is checked, the container does not block movement of the contained DP.

equidistant from any mover.[57] In particular, multiple specifiers of the same projections are equidistant. Thus, minimality is relaxed for elements within the same MDs. Why? Why should elements in the same MD be excused from minimality requirements? This question becomes more pressing when one considers how MDs are defined.

Chomsky (1995a) suggests that elements in the same immediate maximal projection are equidistant for purposes of minimality. But, given conventional assumptions, these specifiers are in c-command configurations, with one c-commanding the other, so why is it that these c-command relations can be ignored for purposes of minimality while others cannot be? This looks like brazen ad hoc stipulation at its worst. Subsequent research has labored mightily to remove this stain.[58] Interestingly, one consequence of the present analysis is that the observed exceptions to minimality immediately follow. Or to put the same point positively, the indicated "exceptions" are not exceptions at all. Consider why.

The standard analysis treats multiple specifiers are equidistant from a given target.

(37) $[_{TP}$ T . . . $[_{BP}$ XP $[_{B'}$ YP [B . . .]]]]

In (37), assume that T is the target. Note the paths of XP and YP are identical; {TP, BP}. Thus, both are equidistant from T. Similarly in (38), movement of XP to α is no longer than movement to β as the paths of the two movements are identical.

[57] See Chomsky (1995a: 356–7, (189)/(190)):

(189) γ and β are equidistant from α if γ and β are in the same minimal domain (190) β is *closer* to K than α unless β is in the same minimal domain as (a) τ [the target, NH] or (b) α.

As Chomsky (1995a) notes:

We thus have two cases to consider. We ask (case (190a)) whether β and τ are equidistant from α, and (case (190b)) whether β and α are equidistant from τ. If either is true, then β does not bar raising α to τ.

[58] See Chomsky (2004) where minimality restrictions are computed at the phase on constructed outputs so as to finesse this problem. Thus, on this conception minimality is a condition on representations rather than one on derivational operations. In my view, this removes most of the computational rationale for minimality and renders minimality restrictions requirements imposed by the interpretive interfaces. However, why the interfaces should impose such a shortness requirement is quite mysterious. In other words, though one can understand why grammatical operations might be computationally better off if they minimized dependency length, it is not clear why interpretive components would so act. One last observation: that dependency length should be minimized has obvious interpretations for performance systems (parsing and production) and the benefit of minimizing such dependency length has natural attractions. It would not be surprising if optimal performance systems and optimal grammars used similar evaluation metrics to measure complexity. For some discussion see Boeckx and Hornstein (2007) and Chapter 7.

(38) $[_{TP}\ \alpha\ ZP\ \beta\ T \ldots [_{BP}\ XP\ [B \ldots]]]$

In other words, XP's movement to the outside edge of TP is no longer than a move that tucks XP in right next to T. The paths are identical in both cases: {BP, TP}. Thus, if we measure distance in terms of paths defined as in (33) above, we derive that minimality effects should not arise between multiple specifiers of a common head as they are elements in the same domain. Such elements always traverse equivalent paths. Note that this is true whether we look at landing sites or launch sites or both. The conception outlined here folds both cases into one. The two configurations define equivalent paths and so the relevant moves are equidistant.

Seen from the current perspective, the theoretical "problem" of explaining why elements in the same MD are exempt from minimality actually stems from taking c-command as a primitive notion. If one defines minimality in terms of c-command then why c-commanding elements within a minimal domain should be exempt is theoretically puzzling. However, once c-command is seen as a derived notion, parasitic on a specification of shortness in terms of paths, it is clear why elements in the same domain should be exempt from minimality. They result in identical paths and so are equidistant from any element outside their common domain.[59]

2.5.3 *Tucking-in and grammatical distance*

Consider a further consequence of the current proposal. As noted, targets in the domain of the same head (even those in c-command configurations) will be equidistant from expressions moving to that target. This suggests that Tucking-in cannot be defined in terms of shortest move. For example, Richards (2001) and Chomsky (2001) suggest that Tucking-in right next to the head involves a shorter movement than merging to a position outside a present specifier. However, if we plot distance by paths as in (33) this is incorrect. Given the problems with Tucking-in in connection with linearization (see Section 2), this is not necessarily an unwelcome result. The Extension Condition requires Merge to apply at the root. Tucking-in is motivated on the assumption that a movement with a merger to the root is longer than one that merges right next to the head. But this requires taking c-command as a primitive, or, more correctly, if one adopts paths as defined in (33) as the measure of distance then the claim that Tucking-in involves a shorter move, though intuitively plausible, has no theoretical standing. More precisely, the intuition that Tucking-in involves a shorter move than merging at the root is anchored in a conception of grammar

[59] The careful reader will have noticed that this reasoning requires computing paths in terms of maximal projections. See note 61.

in which c-command is a primitive relation. If this is questioned, as we are doing here, the utility of the intuition disappears.

We can actually make a somewhat stronger claim. We could amend (33) to redeem the intuition behind Tucking-in. Here's how: simply define a path in terms of *all* dominating projections, not just maximal ones. With paths so understood, moves that tuck-in would traverse shorter paths than those that extend targets. (39) illustrates this.

(39) $[_{TP}\ \alpha\ ZP\ [_{T'}\ \beta\ T \ldots [_{BP}\ XP\ [B \ldots]]]]$

If all projections of T are used to determine a path, then T′ must be included in the path of XP to T in (39). So calculated, the path to β is shorter than the one to α by at least one node, T′.

So, it is possible to accommodate Tucking-in within a path-based conception. But there is a down side to this proposal. It requires adopting a disjunctive definition of distance for we do *not* want to include non-maximal projections in computing the distance for movers (rather than targets). To see this consider (37), repeated here.

(37) $[_{TP}\ T \ldots [_{BP}\ XP\ [_{B'}\ YP\ [B \ldots]]]]$

If we include non-maximal projections in our calculation of paths then because B′ dominates YP but not XP the movement of YP to T should be blocked by minimality. Getting around this problem requires adopting one definition of distance for targets and another for launch sites; in other words, a disjunctive definition of minimal distance. Disjunctive accounts are always methodologically undesirable. However, in the present context such a move has even less to recommend it given the problems that Tucking-in presents for a non c-command-based account of linearization discussed in 2.3. In fact, theoretically, the inability to state the Tucking-in intuition fits well with the idea that it should not be a permissible grammatical option. Of course, this also comes with an empirical promissory note, viz. to reanalyze the data that motivated Tucking-in.[60]

In sum, the methodological problem noted by defining paths in terms of all projections does not arise if we assume the definition in (33) which defines paths in terms of XPs.[61] Curiously, this comports well with the occasionally stated

[60] Hornstein (2001: 144–148) reanalyzes the superiority data from Richards (2001). There remain other cases of interest however. Cf. Chapter 5: Appendix for a discussion of a conception of movement consistent with Tucking-in, though not requiring it.

[61] Chapter 3 shows that the stipulation that paths be defined in terms of maxPs is eliminated if one adopts a strict interpretation of Bare Phrase Structure. On this strict interpretation, the proposed definition of paths above, which accommodates Tucking-in, cannot be stated as it relies on non Bare Phrase Structure notions in defining projections. If this is correct, this provides further support for the present conclusion that Tucking-in is not a grammatical option, though see Chapter 5: Appendix for an alternative conception.

intuition that X′ projections, unlike X^0s and XPs, should be grammatically inert. The two main grammatical uses for X′s lie in stating c-command requirements on binding and measuring distance for tucking-in operations. However, if, as assumed in Section 2.2, binding actually reduces to conditions on overt movement (like Extension, which, as noted, is incompatible with Tucking-in), and Tucking-in is removed as a grammatical option, then these particular motivations for X′s are removed.

2.5.4 Labels and m-command

Paths conceived as in (33) appear to require that phrases be labeled; a path being the union of XPs that dominate the target and the mover (or, the probe and the goal).[62] It is hard to see how this conception can survive in a label free grammar of the kind urged in Collins (2002). Whether this is fatal to the conception is a topic too large to be addressed here.[63] However, it is worth noting that Collins (2002) requires taking c-command as primitive in its probe-goal reanalysis of many of the locality conditions that the X′-account of phrases provided. The reason for this is that, without labels, it is hard to see how domination relations can be grammatically exploited, as we have done here in the definition of a path. (Why? Because though a head may c-command another element in a phrase marker, it cannot dominate it. To state domination relations, labels are very useful.) If so, then there would seem to be a fundamental incompatibility between a label free approach to phrase structure and a project of explaining/deducing c-command like the one outlined here. This need not be a bad thing, however, for it suggests that these conflicting visions actually stem from two different underlying conceptions of locality and phrase structure, both of which have a pedigree in the recent generative tradition (and both of which deserve further investigation). Let me explain.

Once upon a time, there were two conceptions of government and "c-command." One was geometrical and is encapsulated in the definitions in (38).

(38) α c-commands β iff the first branching category that dominates α dominates β. α governs β iff α c-commands β and β c-commands α.

(38) is geometrical because it defines c-command and government in terms of tree configurations, adverting as it does to "first branching categories." For these definitions, the categorical values of the nodes are irrelevant, which is why (38) fits well with conceptions that eschew phrasal labeling.

[62] The next chapter discusses labeling more extensively, arguing that labeling is the quintessential grammatical operation. However, it also proposes that bar-levels should be entirely eliminated (qua targets of operations) from the grammatical computational system.

[63] For a critique of Collins (2002), see Irurtzun (2007).

A second definition, proposed by Aoun and Sportiche (1983), takes a rather different tack. In this approach, the notion "phrasal projection of a head" is central to determining linguistically relevant relations and domains. M-command is defined in terms of XPs which are the maximal phrasal projections of an X^0, a head.

(39) α m-commands β iff every XP that dominates α dominates β. α governs β iff α m-commands β and β m-commands α.

M-command is a grammatically substantive definition in that it relies on domains defined by grammatical objects, viz. phrases. So, while c-command is geometrical in spirit (the locality and domain concepts are graph-theoretic), m-command treats domains and locality in terms of the organization of phrases understood in X′ terms as projections of heads (with the primary cut being among elements within and without the phrasal projection of a head).[64] The path conception developed above has m-command as its intuitive ancestor, which is why XPs are central to measuring distance and why elements within the orbits of the same XPs are not distinguished from one another with respect to relative proximity. C-command cuts more finely than m-command for there often is hierarchical organization within the projection of a head (especially if phrases must have binary branching as typically assumed). The relevant empirical question is whether this additional structure is grammatically relevant. The c-command definition bets that it is; that grammars are sensitive to intra-phrasal hierarchy. The m-command conception bets that it isn't; that grammars regulate *inter* phrasal commerce, not grammatical transactions within a phrasal projection.[65]

What is worth noting for present purposes is not whether m-command or c-command is the correct conception, though this is an important question, but how the various conceptions of hierarchy (graph-based versus substantive) lead to different conceptions of locality and, in particular, how a path-based conception of distance seems to carry a commitment to labeled phrase markers and to an m-command conception of locality.

2.5.5 Sidewards movement (SWM)

There is another consequence of the path-based conception of distance that is worth noting. It applies uniformly to movement within a connected phrase

[64] The graph theoretical intuition is most fully embraced in Kayne's (1984) book on connectedness and binary branching. The substantive intuition finds a plausible home in Chomsky's (1995a) conception of checking and complement domains.

[65] The recent proposal that intra-phrasal movement/remerger is illicit fits well with such an m-command sensibility. See Abels (2003), Bošković (1994), Grohmann (2003), Murasugi and Saito (1995) and Pesetsky and Torrego (2001) among many others for the idea that movements that are too short may be grammatically illicit.

marker and movement between unconnected phrase markers (so-called sidewards or inter-arboreal movement). Let's see that this is indeed so. Consider the two derivations in (40).

(40) a. $[_{vP}\ v\ [_{VP}\ V\ [_{TP}\ DP1 \ldots]]]$
b. $[_{vP}\ v\ [_{VP}\ V\ DP2\ [_{TP}\ DP1 \ldots]]]$
c. $[_{vP}\ v\ [_{VP}\ V\ DP2]]\ [_{TP}\ DP1 \ldots]$
d. $[_{vP}\ v\ [_{VP}\ V\ DP2]]\ [_{TP}\ DP1 \ldots [_{XP} \ldots DP2 \ldots]]$

Consider the movement of DP1 to Spec *v*. This movement targets *v*. The path in (40a) of this movement is {TP, VP, *v*P}. This path describes a movement associated with subject control; examples like *John hoped to see Mary*. The movement of DP1 in (40b) is blocked by shortest move as the movement to Spec *v* is blocked by DP2. The path from DP2 to *v* is {*v*P, VP} while the one for DP1 to *v* is {*v*P, VP, TP}. Thus minimality blocks this move and this is why we cannot get the subject control reading in sentences like *John persuaded Bill PRO (= John) to see Mary*. (40c) is the interesting case. Note that here is a case of SWM as the TP and *v*P are unconnected. The movement of DP1 to *v* yields the path {*v*P, TP}. Note that the path of DP2 to *v* is {*v*P, VP}. Neither is a subset of the other so minimality does not apply to block the movement of DP1 to Spec *v*.[66] This movement underlies cases of adjunct control such as *John saw Mary before leaving the party*.[67] Finally consider (40d), where DP1 and DP2 are both inside the adjunct and DP1 c-commands DP2. Here minimality is expected to block sidewards movement of DP2 over DP1. The path of the former is {XP, TP, *v*P} while that of the latter is {TP, *v*P}. This could prevent sidewards instances of super-raising (**John saw Mary without*

[66] Recall that there are several possible definitions of a path, the one in (33) above and the one in note 48, to name two. In the case of SWM the two define different paths. We have noted how paths are defined for the definition (33) in the text above. However, as applied in the case of (40c), the discussion has tacitly assumed that the target of movement is *v*. This is why *v*P is included in the path as it dominates the target *v*. However, consider a slightly different case: say that the relevant verb were unaccusative so that the relevant structure were (i) (where V and the TP adjunct do not form a connected sub-tree).

(i) $V\ [_{TP}\ DP1 \ldots]$

The path of DP1 targeting V is {TP} by the definition in (33) if we assume that domination is not reflexive. The reason is that V does not dominate itself nor does any projection of V dominate TP. The path given in note 48 is {V, TP} as it is defined to expressly include the target in every path. Note that V in (i) is maximal given BPS as it is unconnected, so we could just as well have written {VP, TP}. It is not clear if anything is wrong with either definition. However, it is worth observing that they produce different results in cases such as this.

[67] Sidewards movement analyses have also been proposed for parasitic gap constructions and ATB movements in coordinate structures. See Hornstein (2001) and Boeckx, Hornstein and Nunes (forthcoming) for a fuller discussion of the adjunct control cases. See Nunes (2004) for discussion of parasitic gaps and ATBs.

it being told <John> that she left) just as minimality blocks this in cases of regular super-raising (**John seems that it was told that she left*).[68]

In sum, the same measure of distance that applies to regulate regular intra-arboreal movement applies successfully to the inter-arboreal or SWM variety. This suggests that these two movements are not really different in *kind*. History, not logic, distinguishes them. As emphasized above (see note 18), both are simply combinations of Copy plus Merge. We see here that both describe identical kinds of paths, measured and compared in the same way. Consequently, on the conception of movement presented here, there appear to be no theoretical reasons for taking sidewards movement to be a novel species of movement any more than the movement involved in Raising is a different kind of A-movement than the one involved in Passive.

2.5.6 *Shortest Attract versus Shortest Move*

There are two different extant conceptions of minimality. One compares movements with respect to a given target; call this Shortest Attract (SA). The other, Rizzi's original proposal, compares the movement of a DP to two different targets; this is Shortest Move (SM). (33) has been stated in terms of SA. However, the approach here is also consistent with an SM account. To see this, consider (41).

(41) $[_A$ T(arget) . . .$[_B$ . . .DP1 . . .$[_C$ . . .DP2 . . .]]]

Say that in (41) T, DP1, and DP2 are in a c-command configuration. Say that DP1, DP2 and T all have the same features. For SA the relevant paths are the ones for DP1 and DP2 with respect to T; P(DP1) = {A, B} and P(DP2) = {A, B, C}. The inability of DP2 to target T in this case is reflected in the fact that P(DP2) is a superset of P(DP1). For SM the relevant paths are different. The idea is that DP2 is blocked from moving to T because it could have taken a shorter move to the position of DP1, a position which has identical features. The relevant paths under this conception are the path from DP1 to T and the one from DP2 to DP1. These are {A, B, C} and {B, C} respectively. Once

[68] There are other ways of blocking these sorts of unacceptable sentences. Hornstein (2006) does so in terms of the costly nature of pronominal use over movement. Thus, the above sentences are ungrammatical because the following sentences are fine: *John saw Mary without being told that she left*, *It seems that John was told that she left*. A phase-based account could also block these derivations, were the clause containing the mover a phase. That one of these, rather than minimality, is actually the relevant condition is suggested by the fact that *it* has checked its case and so should not be a relevant intervener, especially in the second example where the target is a case position (see discussion of the A-over-A above). In the SWM case, the target is a thematic position and so the fact that *it* is case marked need not be as relevant. It will depend on the status of the principle that case freezes movement.

again, minimality is reflected in the fact that the path to T is a superset of the one to DP1.

For current concerns, it does not matter whether one construes minimality as SA or SM. Both result in paths with the right set theoretic properties; with longer paths (i.e. supersets) being blocked by shorter ones (i.e. subsets). This does not imply that both conceptions are equally worthy theoretically or empirically. It only means that whichever is correct is amenable to an analysis in the terms outlined above.[69]

2.5.7 *Merge Over Move*

If we assume that grammars like to minimize dependency length (measured in path terms), it is possible to see the preference for Merge Over Move (MOM) as an instance of this preference.[70] Consider a case in which A is the target of a grammatical operation and either B (and element in the numeration) or C can satisfy its demands.

(42) Numeration = {. . .B . . .}
Derivation: $[_{AP}$ A $[_{XP}$ . . .X . . .C . . .]]

The path of C to A is (at least) {AP, XP}. What's the path from B to A? It is the union of maxPs that dominate the target or the "mover." But in this second case nothing is "moving," so it is more accurate to say that the "mover" is simply the element that merges with A, the "mergee." This is B in the case at hand. Its path is {AP}. Note that this path is clearly a subset of the former. Thus, "pure merge" of B involves a shorter path than "moving" C.

What if AP, rather than A, is the target of the operation (this, after all is where B or C adjoin)? On the most natural construal, domination is a non-reflexive

69 See Chapter 6 for arguments against Attract-based conceptions of UG.

70 The empirical evidence that MOM is a principle of grammar is *not* overwhelming. To my knowledge, there are three known empirical arguments for MOM to date, all of which would be controversial. The most well-discussed case is the unacceptability of sentences like (i) in English existential constructions.

(i) *There seems a man to be here

These cases, however, have been analyzed in a variety of ways which do not require assuming that MOM holds of grammatical computations (see for example the analysis in Chapter 6 below). The second case is the fact that in languages like Icelandic with object shift, if the subject is not moved from its base position, the order one finds is [obj [subj . . .]] and not [subj [obj . . .]]. However, the data cited is considered controversial (especially given the other general assumption that subjects always vacate the *v*P). Another instance involves sidewards movement out of adjuncts in control structures to derive the generalization that adjunct control is limited to subjects (see Hornstein 2001). This too is controversial given its reliance on the existence of SWM. The last case I know of that crucially uses MOM involves Long Distance Anaphora. See Motomura (2001) and McKeown (2007) for discussion. Thus, there are no clearly uncontroversial instances of MOM. This said, the following assumes that it is correct (or at least, might be so).

relation: a node does not dominate itself. If so, the path for the "pure merge" of B and AP above is the null set Ø; nothing dominates B and nothing dominates AP. In standard cases, it makes more sense (at least to me) to see A as the target and not AP, as it is a feature of A that is driving the operation.[71] Interestingly, in cases where simplex lexical items are merged, say a pronoun and a verb as in *saw him*, the path is {Ø} (again if domination is irreflexive and we use the definition in (33)). Nothing dominates either element and so the path must be Ø.

If one further assumes that Merge is asymmetric (B merges with A in (42) above) and assumes that the head that labels the projection to be the target of Merge (A in (42) above), then there seems to be a path-size difference between merging a complement and merging a non-complement to a head. The non-complement merge will always be a superset of the complement merge. This provides a path-based account for the oft made assumption that internal arguments are "closer" to heads than "external" ones are, and implies that complements should always be merged before specifiers (a requirement that falls out from Extension as well).[72]

In sum, if we measure distance in terms of paths we can derive that (pure) Merge always involves a shorter path than any instance of Move. The reason is that (pure) merging any two expressions will involve a shorter path as nothing dominates the mergee if taken from the numeration (or directly from the lexicon). Interestingly, if one assumes that to check a given requirement, shorter paths are chosen over longer ones at any given point in the derivation, then one derives the (possible) fact that Merge trumps Move.[73] It is not clear whether this "derivation" of MOM is a good result or not. For the nonce, I leave it as a curious observation.

2.6 Conclusion

One aim of minimalism is to try and understand why the properties of UG are the way they are. There are two useful strategies for answering this sort of why question.

[71] In Chapter 3 we shall see that the distinction between targeting A versus AP is inert. In effect, the operation of targeting XP is the same as targeting X. If this is so, all cases of pure merge will have null path lengths.

[72] The distinction between merge of complements and specifiers fails if we take the target of Merge to be AP and not A in (42). This is the effect of the proposal in Chapter 3 and so the present observations do not survive the reinterpretation of phrase structure provided there.

[73] There are other derivations of MOM. Hornstein, Nunes and Grohmann (2005) review the other standard way of deducing this; viz. that if Move involves Copy and Merge then it involves a proper subset of the relevant operations for pure merge. Thus if at any given point operations are minimized, Merge should be preferred to Move as it is simpler in a straightforward sense (though see Chapter 7 for some critical discussion). This is not incompatible with the observations above. A good next question is whether MOM holds at all and if so whether the two ways of deriving it are equally useful or actually different. I save this for future inquiry.

One is to reduce what appear to be disparate looking phenomena to common underlying principles. An example of this is the reduction of control and binding to the theory of movement: why must reflexives and OC PROs have local c-commanding antecedents? Because they are formed by A-movement! They have the same properties because they are formed in the same way, by the same rules, subject to the same conditions.

A second is to see the relevant features as the by-products of natural computational principles operative in the domain of language; shorter moves are preferred to longer ones, computations are monotonic, the features relevant for operations are prominent and easy to find (e.g. by being clustered at the tops of phrases): Why in *John persuaded Harry to leave* is the leaver understood to be Harry and not John? Because control configurations are formed by movement and movement is subject to Shortest Move/Attract.

This chapter has combined these two approaches to answer the following question: why does FL/UG use c-command so extensively? The answer provided is that it is what one expects from grammars organized in certain ways. If grammatical dependencies are coded via Merge, if Move is Copy plus Merge (or Merge/ReMerge), if computations are monotonic increasing (i.e. obey Extension), if grammars optimize by preferring shorter relations to longer ones, if grammars use Boolean resources, if grammatical operations are last resort (deterministic), if grammars segregate theta, case and A′ domains, then c-command will figure prominently in grammatical processes. In other words, c-command is what one gets from a well-constructed grammar of a certain kind. The general properties enumerated above are nice ones. Merge and Copy (or Merge and ReMerge) are as basic as computational operations can be. Monotonicity (Extension), non-counting, last-resort, and a preference for short dependencies are all reasonable *general* computational principles. Taken together, this suggests that c-command is what emerges in a system that uses simple natural primitive operations, and deploys them in a computationally optimal way. In other words, perhaps c-command is a mark of optimal grammatical design. Wouldn't that be nice!

3 Labels, recursion and movement

3.1 Introduction: Merge and Concatenate

No introductory course in linguistics is complete without the observation that linguistic objects – words, phrases, sentences – are made up of elements combined in a hierarchical fashion. "Words in a sentence are *not* like beads on a string!" we announce. Words in a sentence nest; they are not a simple linear concatenation of elements. More technically, linguistic objects display recursive embedding, not a simple linear order; sentences have tops and bottoms as well as lefts and rights.

One of the central tasks of modern linguistics has been to characterize the nature of this nesting. Within syntax, the consensus opinion is that recursion is the province of the phrase structure component of the grammar. Here rules can apply repeatedly without limit to nest categories within one another. As it is possible to embed a category of type X within a category of the same type, recursion emerges and hierarchically nested structures emerge without limit. The recursive trick is encapsulated in (1), where a phrase of type X is contained within a phrase of type X.

(1) $[_{XP} \ldots XP \ldots]$

The varying generative accounts of phrase structure all allow structures like (1). Theories of the *Aspects* vintage generate such structures using rules like (2).

(2) a. S → NP VP
 b. VP → V (NP) (S)
 c. NP → Det N (PP) (S)
 d. PP → P NP

Applying (2a) and (2b) will allow an S to be embedded within an S recursively. (2c,d) will allow NPs to be recursively generated within NPs without limit.

All other generative approaches achieve the same ends, albeit with slightly different means. In GB, phrase structure rules are streamlined along the lines of X′-theory; phrases being understood as projections of lexical heads with a basic structure as in (3), with ZP and YP being specifiers and complements of the head respectively. The phrasal character of ZP and YP allow for further expansion meeting the template in (3), and this grounds unending embedding.[1]

(3) $[_{XP}$ ZP $[_{X'}$ X^0 YP]]

Minimalist accounts return to an earlier view of phrase structure. Phrase structure rules are replaced by a Merge operation. The idea is that lexical items are culled from the lexicon and combined to form special kinds of sets. The idea is made clear with an example. Consider the sentence in (4).

(4) John likes the dog

It is made up of the words *John*, *likes*, *the*, *dog* and various functional elements like tense. These words are combined to form sentences as follows.

(5) a. Merge *the* and *dog* → {the,dog}[2]
b. Merge *likes* and {the,dog} → {likes {the,dog}}
c. Merge *John* with {likes {the,dog}} → {John {likes {the,dog}}}
d. Merge T(ense) and {John {likes {the,dog}}} → {T {John {likes {the,dog}}}}
e. Copy *John* and Merge it and {T {John {likes {the,dog}}}} → {John {T {John {likes {the,dog}}}}}

There are various operations that we have abstracted away from here that we need not clarify at this point. Let's take a look at (5) above. Merge repeatedly applies to provide bigger and bigger structures. Note that it applies both to atoms culled from the lexicon and to molecule-like outputs of prior Merge operations. Note too that it applies (by stipulation) at the "root." These two assumptions (that Merge applies at the root and that it applies indifferently to atoms and "molecules") are what yield the hierarchical structure of phrases and sentences. As we keep merging in (5) above, we construct a set with "deeper" subsets, i.e. we induce a hierarchical nested arrangement. The naturalness of these two assumptions should not obscure their axiomatic nature. Merge need not have been so defined. There is nothing incoherent about another operation, call it Merge′, which operates to yield a flat structure as in (6).

(6) Merge′ {A} and {B, C}→ {A, B, C}

[1] Adjunction is another productive recursion generating device. Phrases can be adjoined to XPs to produce phrases of the same XP category.

(i) $[_{XP}$ $[_{XP}$ XP] YP]

YP can be on the left or the right. What is important for purposes of unbounded recursion is that the output of adjunction returns a category of the same type as one of the inputs. For the present, we will put adjunction to one side. For more detailed discussion, see Chapter 4.

[2] Note Merge is thus similar to a comprehension operation within set theory; elements are rounded up and put into a set.

Merge′ is an operation which will output a string of non-nested elements. Atoms are identified with their singleton sets and Merge′ is identical to set union. Observe that it is recursive. There is no upper bound on how many distinct elements the operation can assemble into a set.[3] Similarly, the operation Concatenate as in (7) can construct longer and longer strings without any nesting.

(7) Concatenate: A, B → A^B : C, A^B→ C^A^B

Thus, the distinctive characteristic of UG is not that grammars can generate longer and longer structures but that these longer and longer structures have a nested structure.

A tacit assumption in discussions of recursion in UG is that Merge and Merge′ or Merge and Concatenate are entirely distinct operations and that Merge is unique to UG. This chapter proposes that Merge is actually a species of concatenation. Or put another way: on the assumption that concatenation is a more primitive operation than Merge, I want to consider what must be added to a concatenative system to yield Merge.[4] I will suggest that *labeling, understood as it is within Bare Phrase Structure,* supplies the necessary ingredient to get one from a flat beads-on-a-string system to a hierarchical nesting system. Thus, if Merge is a species of concatenation, then labeling (in particular endocentric labeling, an operation that renders its output as type identical to one of its inputs) is the central innovation of UG, the change that enables the peculiar architecture of natural language to emerge. Moreover, the labeling that induces hierarchy brings in its train two further grammatical properties displayed by natural languages (NL); the Endocentricity Restriction (the fact that only the head of a maximal projection is visible from outside that

[3] There is an upper bound if Merge′ is interpreted in terms of set creation as the set {x, x} = {x}. However, if we assume that each *choice* of lexical atom counts as a distinct member (e.g. with a separate index as is assumed when numerations with multiple expressions of the same item are assembled as in the sentence *The dog saw the dog* there are two "the"s and two "dog"s), then there is no upper bound at all.

[4] Concatenation is also discussed in Epstein (1999). An important caveat about how "concatenate" is used here: I abstract from the fact that concatenation imposes an order on the affected elements. In this book, I assume that one can have order free concatenation so that A^B and B^A are indistinguishable. One can think of this as concatenation in a 2-space, where the concatenation of A^B does not determine whether A precedes or follows B. Perhaps a better name for the operation might be COMBINE. However, it is the practice to differentiate linguistic from concatenative systems, so sticking to "concatenate" will help keep this contrast in mind. What is critical in what follows is not the name of the operation but (i) that it pick out that operation that antedates the one that the faculty of language uses to generate the recursively embedded structures typically found in natural language, and (ii) that it typically delivers non-hierarchically ordered (flat) objects.

Last point: it is currently fashionable to distinguish operations that specify hierarchical relations from those that linearize these structures via some algorithm. However, this was not always so. In the earliest days of generative grammar, PS rules determined both hierarchical and linear order. It is not obvious what goes wrong if we return to this assumption. However, in what follows I will assume that our departure from these early approaches is correct and that Concatenate does not specify a linear order.

projection) and displacement (aka movement). Thus, Labeling in conjunction with other *non*-linguistic cognitive operations (in particular, concatenation and copying) suffices to yield generative systems with three of the most distinctive characteristics of NL grammars, or so I will suggest. If this is correct, then the evolutionary "gateway" innovation that made natural language possible might reduce to one rather trivial operation, (endocentric) labeling, which in concert with other cognitive operations already in place flowered into the faculty of language (FL).[5] A pleasant consequence of this view, if sustainable, is that it might help account for the rather rapid emergence of language in humans discussed in Chapter 1.

Here's the game plan for what follows. We start with a fussy discussion about what concatenation is and what it presupposes to be well defined. We next ask what labeling is and how it interacts with concatenation to deliver hierarchy. We understand labeling in the traditional sense (cf. Chomsky 1957, Lasnik and Kupin 1977). It is the technical device that underlies the *is-a* relation. Endocentricity is that species of labeling in which one of the inputs serves to type specify the concatenated output. If labels are "bare" (in the sense of Bare Phrase Structure (BPS), Chomsky 1995a, b) then endocentric labeling functions to turn concatenated atoms into complex atoms and hence liable to further concatenation. Such labeling suffices for hierarchical recursion to emerge, as well as other common features of natural languages (e.g. endocentricity restrictions). Moreover, in combination with other *pre*-linguistic operations like Copy and Concatenate, the system yields displacement (aka, movement). We further note that if these computations are computationally well-behaved (i.e. respect a principle like Shortest Move understood as minimizing path length as in Chapter 2) then the system shows other properties distinctive of natural languages (e.g. structure preservation, constituent movement). This clearly has implications for how the operations underlying natural language structure might have emerged and we end with some brief discussion of these concerns.

3.2 What is concatenation?

Concatenation is the most elementary mode of combination[6] and, like all operations, it is defined over a set of atoms.[7] This point is important so I will

[5] Labeling then would constitute the main feature of the Narrow Faculty of Language in the sense of Hauser, Chomsky and Fitch (2002).

[6] It is "most elementary" not in the sense of being conceptually the simplest (whatever that might mean), but in being more cognitively primitive; i.e. we consider stringing together atoms non-hierarchically as "simpler" than doing so hierarchically.

[7] This conception of concatenation is present in Chomsky's (1955) conception of a "level." A level is defined as concatenation over a set of primes. Levels do not mix and they are distinguished by their differing primes. This is analogous to the assumption made here, viz. that concatenation implicitly comes with a set of primes/atoms over which it operates.

belabor it. What one gets via concatenation depends on what atoms one is manipulating. Concatenating the letters *t*, *h*, *e*, *c*, *a*, *t*, can yield, among others, the concatenative complexes t^h^e^c^a^t or t^c^h^a^e^t while concatenating the atoms *the*, *cat*, yields the complex the^cat and cat^the. These complexes are weakly similar (thus t^h^e^c^a^t and the^cat have the same string order) but strongly distinct (thus t^c^h^a^e^t is a possible output of concatenating letters but not of concatenating words).[8] In short, what atoms concatenate is critical to determining the complexes that can be formed.

Concatenation is a very promiscuous operation. Its atoms can include phonemes, letters, syllables, words, sentences, actions, plans, flowers (think daisy chains), whatever. It is a virtual certainty that non-verbal beings can concatenate some elements (though not others) and string them together into larger ordered objects. In this sense, concatenation is not a linguistically specific operation, i.e. one restricted to the faculty of language.

It is regularly assumed that concatenation is *not* the operation that knits the atoms of a sentence together. Why? For the following reason: If we assume that concatenation is defined over atoms (i.e. that only atoms concatenate) and that the atoms which combine to form a sentence are words/morphemes, then concatenating them can only result in flat beads-on-a-string structures. Thus, concatenating A,B,C, as in (7) above yields the flat A^B^C and concatenating *the*, *dog*, *barks* yields the^dog^barks and not our familiar and beloved [[the dog] barks]. Thus, we conclude that it cannot be that sentences are concatenates of words (atoms), since sentences are hierarchically structured.

Though this has a convincing ring to it, the reasoning relies on a hidden premise of note: viz. that concatenates (i.e. A^B) are not atomic, though their parts, i.e. A and B, are. If we give this assumption up, then concatenation alone suffices for constructing the nested dependencies seen in sentences. In other words, *if* previous concatenates can be inputs to further concatenations, then concatenation yields hierarchy. Thus, [the^dog]^barks is hierarchical and what makes it so is that the prior concatenate the^dog is input to the next concatenation with *barks*. At the risk of being pedantic, what distinguishes the^dog^barks from [the^dog]^barks is that in the former *barks* concatenates with *dog* while in the latter it concatenates with *the^dog*, the latter being treated as if it were an atom subject to concatenation. So, if concatenation can treat its (complex) outputs as (atomic) inputs liable to further concatenation, hierarchy emerges.[9] The important linguistic question, then, is what licenses taking a constructed concatenate as input to further concatenation given *(i) that the*

[8] The terms "weak" and "strong" are used in the sense familiar from Chomsky (1965).

[9] Chomsky (1955) considers a similar issue: when is the output of a transformation a possible input for a transformation? It cannot be taken for granted that the output of an operation on some primes can serve as input to these same operations. Chomsky's discussion of "edge features" in more recent work addresses this same issue (see Chomsky 2005b).

concatenation operation is defined over atoms and (ii) that a (constructed) concatenate is not itself an atom? The answer: Labels. Let's see how.

Chomsky (1995a,b) analyzes phrase building as consisting of two operations. The first (surveyed in (5)), Merge, takes a pair of atoms and combines them. If Merge is just concatenation, this is the operation that concatenates the two as in (7). The second operation is labeling. This is an operation whereby one of the two inputs to concatenation "names" the resulting concatenate. This is illustrated in (8).

(8) a. Concatenate A,B → A^B
 b. Label A^B → $[_A$ A^B]

The square brackets here should be read as saying that the concatenate A^B has label A.

What does labeling do? Or, what does "$[_A$ A^B]" mean? In Chomsky (1955), labels on phrase markers were understood as defining the "is-a" relation. Thus (9) says that V^NP is a VP.

(9) a. VP → V NP
 b. $[_{VP}$ V NP]

However, as Chomsky (1986a: 18) notes, it is a principle of X′ theory that "a head and its projections share *all* [my emphasis, NH] properties apart from bar-level, including the features involved in Θ-marking." Combining this with Bare Phrase Structure implies that labeling maps a lexical item plus its concatenate back into the lexical item itself (after all, it shares all the properties of A). Thus, given that the lexical item A was a concatenable, so too is $[_A$ A^B]. In other words, what the labeling in (8b) says is that the complex concatenate *is a concatenable atom*; in fact, just (an) A.[10] Thus, the output of this labeling convention is an atom for the purposes of (further) concatenation if we understand labeling as in Chomsky (1955) and endocentric labeling as in Chomsky (1986a).[11]

[10] The formal effect of labeling is to generate a *closure* of concatenation within the domain of lexical atoms by mapping each concatenated complex to one of its atomic parts. As a result, labeling creates an equivalence class of expressions all liable to concatenation. One might thus think of labeling as how a lexical atom generates an equivalence class of structures all subject to the syntactic operations that the lexical item itself is subject to. Pietroski (2007) explores a Fregean analogy, often noted, that may run deep: one can characterize the natural numbers in terms of zero and the relation *less than*, which is the transitive closure or "ancestral" of the predecessor relation.

[11] Boeckx (2006) suggests a mechanism for labeling. It is a species of copying, wherein one of the inputs is copied onto the concatenate. What is critical is that such copying be understood as delivering the "is-a" relation, i.e. where the label categorizes the complex it labels as an instance of the labeling expression. Thus labeling incorporates what Boeckx describes as rigid categorization (dominance by type) and so it is not surprising that when labeling emerges so too does this cognitive ability. See Boeckx for further discussion.

There is a second critical ingredient: we must treat labels in a bare phrase structure (BPS) manner. Note that there are no bar levels on the label in (8b). It is *not* labeled A′ or AP but simply as A. This comports with the BPS idea (following Muysken 1982) that bar-levels are (at most) relational properties of a phrase. There is no *intrinsic* difference between an A, A′ and AP.[12] If the concatenation operation (henceforth: Concatenate) is blind to these relational properties, then as far as this operation is concerned, a labeled concatenate simply is an atom and thus subject to further concatenation. In other words, if there is labeling and we interpret labels as (i) defining the "is-a" relation in a BPS manner, (ii) inheriting all the properties of the head (as in Chomsky 1986a) and (iii) we restrict Concatenate so that it only recognizes the intrinsic features of items (not relational ones), then we allow the derivation of nested structure using a concatenation operation.[13] Or, given labels and the derived atoms that they produce, Merge can be identified with concatenate.

One technical point before proceeding; labels here are not identical to those in Chomsky (1995a,b). The label of a derived structure is one of the atoms that concatenated. Labeling identifies a complex structure with one of its (atomic) inputs. The label is just one of these. In Chomsky (1995a), labels are quite complex. They, in effect, recapitulate the history of derivation. Here, labels are much more similar to the traditional ones in X′-theory; the label being the head that "projects." The more complex conception in Chomsky (1995a) will not serve present aims. The reason is that labels serve to make a concatenative complex atomic via the "is-a" relation. Complex labels cannot do this as they do not denote atomic elements. Thus, labeling must be simple. It is the operation that names the output of a concatenation of atoms for one of the inputs thereby rendering it type identical to one of the atoms.

3.3 The payoff

Is there any value in so construing Merge? I think there is. Let's consider some.

3.3.1 Endocentricity and recursion

First, we can now analyze nested recursion as a function of two operations; one plausibly pre-linguistic and one linguistically innovative. The pre-linguistic operation is Concatenate. It is recursively applicable: given A^B we can derive C^A^B and then A^C^A^B (if A,B,C are atoms). So in answer to the query: why

[12] As noted in the quote from Chomsky (1986a) above, the various projections of X are only distinguished with respect to bar-level. Thus they share all other (intrinsic) properties.

[13] Concatenate must *at least* recognize inherent properties of lexical items for "first" Merge/Concatenate for at this point there are no relational ones. Given this, the null assumption is that *only* inherent features count when applying Merge/Concatenate.

are sentences of unbounded length in natural language? The answer is because concatenation can apply repeatedly. The specifically *linguistic* contribution comes from labeling. It is the source of (unbounded) nesting for it functions to turn a non-atomic concatenate into an atom liable for further concatenation.

Labeling as construed here carries (nested) recursion on its sleeve; from something with an A as part we return an A. This is all that is needed to get recursion going, for recursion obtains within a rule system once a structure of the form $[_{X(P)} \ldots X(P) \ldots]$ can be formed. For example, a TP within a TP or a DP within a DP is sufficient to yield endlessly nested structures. If this is so, the operational source of the kind of nested recursion we find in natural language can be localized in the endocentric labeling operation peculiar to syntactic expressions. To be tendentious, endocentric labeling is the evolutionary innovation which when combined with concatenation yields the unbounded nested structures characteristic of natural language sentences.

Let's put this one more way by contrasting our proposal with what I am not suggesting. I am *not* saying that labeling is a necessary condition for nesting and hierarchy. Logically, hierarchy and recursion are independent of labeling. And not only logically: for example, it has been observed that there is nesting in language without endocentric labeling, as in, for example, syllables which have [onset [nucleus coda]] structure. There is no endocentric labeling in syllables and, interestingly, we do not find repeated nesting in such configurations, i.e. syllables within syllables.[14] Endocentric labeling provides a recursive template and when combined with Concatenate is sufficient for generating unbounded embedded structures. Again, this does *not* imply that recursive embedding is logically contingent on labeling. Early theories of phrase structure treated sentences as non-endocentric (i.e. Aux/Infl was not treated as the head of the sentence). Nonetheless, such systems did have recursive embedding, as a brief look at the rules in (2) makes clear, viz. S nodes expanded to include VPs, which in turn expanded to include S nodes. So, endocentric labeling is not *necessary* for hierarchy or recursive embedding. However, it is *sufficient* for both, and as research over the last 40 years has provided very good evidence that the phrase structure rules of natural language *are* endocentric, I am proposing that this feature brings in its train the kind of unbounded hierarchy characteristic of natural languages. In sum, the proposal ties two facts closely together: First, that endocentricity is unique among the cognitive hierarchies one finds in biology and second, that unbounded hierarchical recursion outside of language is very rare.[15] The current proposal suggests that this is not an accident. Though logically hierarchy and recursion are each possibly independent of endocentric labeling, in fact endocentric labeling in the context of concatenative systems

[14] The common assumption that syllables are hierarchically structured or even that they are linguistically significant units of analysis is currently contentious. For an excellent critical discussion, see Samuels (2008) and references therein, especially Tallerman (2006).

[15] As noted in Boeckx (forthcoming).

suffices to yield the unbounded hierarchical recursive structures of the kind natural languages employ, and assuming that they stem from a common source, it would tie these two distinctive properties of the human language faculty tightly together.[16]

3.3.2 The endocentricity constraint

So construing labeling allows a plausible account for a second distinctive property of grammars. Head-to-head relations (e.g. selection, subcategorization) are restricted to the heads of phrases (let's dub this "the endocentricity constraint").[17] Thus C can select for a finite or non-finite T but it never imposes restrictions on the complement of T (the *v*P/VP) or the specifier of T (the DP subject). Similarly, a higher V may select for a +/−Wh C but it imposes no restrictions on the Spec of C or the nature of TP.[18] Why are these head-to-head relations so restricted? A plausible answer is that all such relations must be established under concatenation. In other words, say that all inter-lexical relations were parasitic on concatenation, then a head X could select/subcategorize Y only if X concatenated with Y. If selection/subcategorization is a species of feature checking (and what cannot be so construed given the labile nature of features) then this becomes the requirement that feature checking is restricted to concatenates. Were this the case (and note this strongly restricts such operations by making them subject to a very local relation), then given that concatenation only holds among atoms, one would expect that in complex concatenates only the head (i.e. label) would be "visible," as the whole labeled expression is atomic for the purposes of concatenation. Thus, in (10), C concatenates with A (recall again, concatenation is between atoms, and labeling here means that A^B is an atom A) and so C can only "see" A in this labeled structure.[19]

(10) a. $C\hat{}[_A A\hat{}B]$
b. $C\hat{}[_A D\hat{}[_A A\hat{}B]]$

16 Rob Chametzky has pointed out to me (p.c.) that this leaves it open whether labeling is a response to some kind of evolutionary pressure. This is correct. It is consistent with what is proposed here that there is a general pressure, say, for hierarchy on, say, computational grounds of the kind Herbert Simon has pointed to. If the particular response to this pressure is endocentric labeling, then the kinds of structures we see in natural language would result. It is also consistent with what is proposed here that such labeling is entirely adventitious (e.g. the result of a genetic mutation) and not related to selection pressures of any kind. Endocentric labeling is the "mechanism" required to get hierarchical recursion in the context of an existing concatenative system. Whether it arose by "selection" or in some other way is an independent question.

17 This is referred to as the "periscope property" in Hornstein, Nunes and Grohmann (2005).

18 For a review see Hornstein, Nunes and Grohmann (2005).

19 See Epstein (1999: 320ff) for a similar idea. Here too Merge is understood as concatenate and grammatical relations are only allowed to form among concatenates. The main addition made here to Epstein (1999) is the reinterpretation of labeling along Bare Phrase Structure lines.

3.3.3 *The Extension Condition*

Third, it follows that Merge must always be at the root. It is generally assumed that first merge must be at the root.[20] Thus, in (10), C cannot merge with B as this would not be merging at the root. One consequence of this assumption is that it prohibits the derivation of structures with multiple mothers (i.e. where in a structure like (10) above, C merges with B and projects a label yielding something like (11) (the indicated lines above A, B, C, represent that A has concatenated with B and C has concatenated with B).

|. . . |. . . |

(11) $[_A$ A $[_C$ B] C]

If we take Merge to be a species of concatenation and if we take labeling to return atoms, then it follows that Merge must be "at the root." Or, to put this another way, because concatenation always applies to atoms it cannot see anything but "roots." In (10a), for example, after the labeling of A^B as A the internal structure of $[_A$ A^B] is invisible to concatenation. The only eligible target of concatenation is the A-labeled structure. It is thus no surprise that this is what Merge targets.

The reasoning is analogous to the following: imagine that lexical items are complexes of features. So an element A is actually $[_A$F1, F2, . . . Fn] and B is $[_B$ G1, G2, . . . Gn]. We need not specify that combining A and B in the syntax yields A^B, viz. $[_A$F1, F2, . . . Fn]^$[_B$ G1, G2, . . . Gn], and not $[_A$F1, $[_B$ F2, . . . Fn, G1], G2, . . . Gn] where A and B overlap. Overlap is blocked on the assumption that A and B are atomic *at this point in the derivation with respect to this rule*. Thus, combining *the* and *cat* must yield the^cat and not th^ceat because *the* and *cat* are syntactic atoms.[21]

The reasoning here is different from that proposed in Chomsky (1995a). Chomsky (1995a) argues that Merge is at the root (i.e. it obeys Extension) because this is the least complex kind of Merge operation. More to the point, merging anywhere but at the root, it is claimed, is far more complex. Chomsky

[20] See Chomsky (1995a) and subsequent writings where this assumption is retained. Recently, some have challenged this assumption. See, e.g., Citko (2003), Wilder (1999). As should be evident from the above, these latter approaches are incompatible with the proposal explored here on the assumption that labeling is strictly cyclic and must apply immediately if it applies at all. Thanks to Tomo Fujii for this observation. However, see Chapter 5: Appendix for a slightly revised set of assumptions that are compatible with multiple domination structures.

[21] See Chomsky (1955) for relevant discussion in the context of levels. Note too that if Merge is the concatenation of atomic elements then the fact that linearization also prohibits overlap between linearized elements might also follow if the linearization algorithm specifies the *direction* of concatenation, e.g. linearization specifies whether the structure is spelled out as "Concatenate left or right." Thus, if linearization presupposes Concatenate, the atomic nature of the constituents must be respected, and the lack of overlap follows.

(1995a) does not go into much detail (it does not demonstrate how much more complex merger at non-roots would be) and when one tries to see what added complexity ensues if merge to a non-root is permitted, it is not clear that the additional complexity is particularly daunting. In fact, given that ReMerge (an instance of Merge given current conceptions) must select elements that are non-root (viz. the remerged elements are non-root), it is unclear why first Merge cannot exploit the very same technology to merge with non-root elements. In other words, if non-root elements can be identified for purposes of movement, why is it so hard to find them for purposes of first Merge?

The present account contrasts with the one in Chomsky (1995a) by being purely formal. It relies on Merge being a species of concatenation. Concatenation is always "at the root" (i.e. obeys Extension) as it is defined to apply between atoms. If labeling turns a complex concatenate into an atom for further concatenation, then it is no surprise that its "internal" structure is not a possible target of concatenation. This is what it is to be an atom: there is no relevant internal structure. Of course, should Chomsky's (1995a) reasoning be correct, then it would reveal a computational virtue of labeling and thus provide a functional rationale for its existence: it allows concatenation to extend to complex structures in a computationally optimal manner. However, this is not to explain *why* Merge is at the root, but *why*, given that it is so, it might be computationally prized.

3.3.4 *The why and what of movement*

If elements impose conditions on each other only under concatenation (i.e. if aRb then a^b),[22] then movement must exist given other features of grammars. Let me elaborate. Say that grammatical relations can be established under concatenation as proposed above. Then, on methodological grounds (given, of course, all the standard caveats about things being equal), we should restrict grammatical interactions *exclusively* to those that can be established under concatenation. If so, if a given element A must enter into several relations with different heads B, C, etc. carrying various kinds of features, then the only way to accomplish this is for A to concatenate with each of the relata A, B, etc. For example, if theta marking requires a relation between a DP and a Verb and nominative case is assigned to a DP when it is related to a finite Tense, then a DP that needs both a theta role and a nominative case must concatenate with both V and T. However, this is only possible if there is some form of

[22] Last Resort/Greed plausibly renders this a biconditional: aRb iff a^b. In other words, if all grammatical operations must be motivated by some kind of grammatical relation then concatenation is only permitted if some relation is established between the concatenates, e.g. theta marking, case checking, agreement, etc.

displacement given the fact that labeling transforms concatenative complexes into atoms for purposes of further concatenation. Let me illustrate.

Say that in (12), B needs to check a feature of C or a feature of itself against C. B cannot concatenate with C where it sits as it is inside an atom, A, and so not available for (is invisible for) further concatenation.

(12) $C\hat{}[_A\ A\hat{}B]$

The only way for B to concatenate with C is as in (13).

(13) $B\hat{}[_C\ C\hat{}[_A A\hat{}B]]$

The derivation in (13) illustrates displacement. B has been copied and concatenated with a complex atom labeled C, i.e. it has merged with C. So, if we require that grammatical relations be established under concatenation and if expressions cannot discharge all of their grammatical obligations by a single application of concatenation, then movement follows from the fact that concatenation is required for grammatical interaction.[23]

Let me elaborate a bit more. The motivation for copying B and concatenating it with C in (13) is the assumption that *only* concatenation can establish grammatical dependency. If, for example, C could interact with B under an operation like AGREE, then there would be no reason to move, for long distance grammatical commerce could be conducted between non-concatenating

[23] The following is an interesting question: why can't single heads bear all the features required to discharge all of a concatenate's grammatical requirements? In GB, for example, an object's case and theta requirements are discharged by V. As such, the object need never move. However, if one assumes that case relevant heads are distinct from the theta relevant ones (as holds in current minimalist accounts where case is due to *v* and theta role to V), then movement will be forced if a DP object is subject to both case and theta requirements. The deeper question, which, to my knowledge, nobody has addressed convincingly, is why heads cannot discharge complex sets of features and why sentences divide roughly into three sectors; a theta domain, a case domain and an A′-domain.

The only seeds of an answer to the first question that I am familiar with is based on speculations implicit in Pollock (1989). He assumes, roughly, that there is a one-to-one correspondence between heads and features. This excludes heads from discharging several features at once (e.g. agreement and case or theta and case) and given that elements have more than one feature to discharge, movement is inevitable. Most current minimalist analyses following Chomsky's (1995a) argument against Agr heads assume that agreement (a complex set of ϕ-features) can be assigned to T and *v*. This appears contrary to Pollock's assumption, though subtle variants are no doubt conceivable.

This leaves the second question: why three domains? Paul Pietroski (p.c.) suggests the following speculation: that the division between case and theta domains functionally supports the movement of DPs and so provides the open sentence (aka: nuclear scope) that are required to interpret quantifiers. On this view, case sub-serves quantification (or, at least quantification via determiners). This idea gains support from the oft-noted observation that in many languages only the strong determiners *require* case; weak indefinites being licensed in other ways (e.g. Turkish as in Enç 1991 and Cagri 2005). This makes sense if part of what case is doing is allowing certain Ds to scope out of their thematic domains to allow their determiners to be interpreted. For some suggestions compatible with this general idea see Diesing (1992) and Hornstein and Uriagereka (2002). See Chapter 7 for further discussion.

elements via AGREE. Restricting grammatical interaction solely to concatenating elements forces movement. This, of course, is not news. Movement in earlier Minimalist models, circa the mid 1990s (Chomsky 1993, 1995a) was driven by the need to check features with feature checking restricted to Spec-X^0 configurations. This account of movement is lost if AGREE is added to the grammatical repertoire of operations. The present proposal returns to the earlier conception. What has been added is the observation that, properly framed, the Endocentricity Condition, nesting, and movement can all be traced to the same assumptions, viz. that grammatical structures and relations rely on concatenation of labeled atoms. In short, *how* nested recursion comes about in natural language brings along movement and the locality restrictions characteristic of endocentric phrases found in natural languages.[24]

Three last points before moving on:

First, the above assumes that Copy contrasts with Concatenate in not being restricted to atoms. In other words, it is licit to copy *part* of an atom (e.g. the left half or the top third or the middle 3/19ths). Thus, in (12) above, B has been copied though B is contained within an A-labeled concatenate. This should be impossible were a labeled element atomic *for the Copy operation*. It is not. What we have proposed above is that labeled elements are atomic *for Concatenate* as this operation is *defined* with respect to a set of atoms. This need not hold for Copy. Anything can be copied, but only atoms can be concatenated. This is a crucial detail (revisited in Chapter 5: Appendix).

Interestingly, Copy alone cannot establish grammatical relations. Only Concatenate can. Importantly, allowing Copy such latitude does not lead to generative profligacy as copies must be re-concatenated for derivations to converge. This assures that in licit derivations only concatenative atoms will be copied as only they can be reintegrated into the structure via Concatenate. This allows Copy to remain unrestricted and thus apply to subatomic parts. Via Copy then, an element buried inside a complex concatenate can enter into further concatenations (further grammatical relations). In other words, together Copy and Concatenate allow an element that is otherwise inaccessible to enter into novel grammatical relations. Displacement, then, the pairing of Copy with Concatenate, is the mechanism by which an expression is able to enter into relations with disparate relata in a system where grammatical relations are only forged via concatenation.[25]

[24] See Chapter 6 for further discussion.

[25] To say that α is atomic is not to say that it has no "insides." Rather it is the claim that α's insides are structurally undifferentiated. Thus, they cannot be the objects of operations like Concatenate which targets differentiated atoms. Copy, however, can apply to an undifferentiated mass. It need not target atoms. Of course, for a copied mass to be further concatenated it must coincide with an atom, as concatenation is only defined over atoms. In sum, that Copy can apply to the insides of an atom is consistent with the claim that α is atomic in the sense of being the smallest structurally differentiated unit.

Second, it is crucial that the Copy operation *not* be (an instance of) Concatenate. Recall, that with respect to Concatenate, B is invisible as it is buried inside a concatenative atom (a complex labeled A). Thus, if Merge is Concatenate, then ReMerge cannot be ReConcatenate. In particular, in (12) above, B within $[_C\ C\hat{}[_A\ A\hat{}B]]$ cannot (re)concatenate with C as it is inside a concatenative atom and so is invisible to further concatenative processes. To become available for further concatenation, we need a copy. This clearly differs from some current conceptions of movement as simply a species of Merge (Internal Merge) and, if on the right track, it provides a rationale for Copy-based approaches to movement and against reducing Move to an instance of Merge.[26,27]

Third, given these assumptions, both first Merge and Move must obey the Extension Condition as both involve instances of Concatenate. A copy can only concatenate if it is an atom, but if it is then it can only concatenate the way other atoms do, i.e. at the root.[28]

3.3.5 *Specifiers and complements*

The current approach explains why Spec-X^0 relations are grammatically ubiquitous and, in some ways, privileged. One reason that recent minimalism has migrated from earlier Spec-X^0 accounts towards AGREE-based conceptions has been the difficulty of conceptually motivating a special role for specifiers (cf. Chomsky 2000). The argument is as follows: Though complements might

26 In my opinion, the brouhaha over whether ReMerge should replace Copy and Merge is largely a tempest in a very small teaspoon (not even a cup!). In most ReMerge accounts the relevant computations are defined over occurrences of an expression, not the expression itself. As these are virtually identical to copies of an expression on a Copy/Merge account, the two conceptions come very close to being notational variants (see Kobele 2006 for a detailed discussion). There are two possible ways of teasing the Copy versus the ReMerge theories apart. One concerns the question of whether multiple copies can receive phonological expression (see Boeckx, Hornstein and Nunes 2007 for some discussion). The other relates to very abstract concerns like the one explored here where we consider the fine details of concatenative systems, and see if Merge is simply one of these. On both counts, I believe, the arguments favor the Copy approach over the ReMerge theory. However, see Chapter 5: Appendix for further discussion of Multi-domination and its relation to Copy.

27 It also raises the question of the source of Copy. Is this also a Faculty of Language innovation or is it a more generally available (and, hence, more primitive) cognitive operation? My own hunch is that it is more likely the latter. Just as Concatenate operates across cognitive modules allowing some cognitive complexity in the absence of generalized nesting, so it seems that animals can string together repeated sequences of the same behavioral atoms (e.g. strings of notes repeated for song, or sequences of actions that include the same parts). If this is so, then Copy is another pre-linguistic cognitive operation that, like Concatenate, can be recruited for linguistic purposes. For some further discussion of the conceptual status of Copy, see Collins (1997) and Hornstein (2001).

28 It is worth observing that any account that incorporates Extension will merge and move only to the "edge." Hence, any such theory, including the present one, "explains" why *if* there is movement it is to an edge and not someplace else. In this sense, the atomicity assumption proposed here for labeling derives the effects of Chomsky's (2005b) "edge features."

enjoy special relations to heads (being *the* internal arguments), specifiers are the "elsewhere" case (i.e. specifiers being the *non* complements). Why checking domains should be restricted to the elsewhere case is conceptually murky and this has weakened the attractiveness of a Spec-X^0 approach to grammatical relations.[29]

The present account goes some way to addressing this conceptual problem. On the view above, grammars establish grammatical relations via concatenation. Concatenation is limited to atoms. The Spec-X^0 relation is simply an instance of concatenation among atoms, one of which is complex, i.e. a labeled concatenate. There is nothing special about the Spec-X^0 structures beyond being the result of a common kind of concatenation. In (13), (viz. $B\hat{}[_C\ C\hat{}[_A A\hat{}B]]$) we can describe B as being a Spec of C. However, the operation has simply concatenated (a copy of) B and C and that is all that is grammatically relevant. What particular grammatical obligations are discharged under concatenation depends on the atoms that are combined. If in (10a), A is a V and B is a DP then the relevant relation is theta marking. In (12), the relation between B and C might be case or ϕ-feature checking or theta marking (if C happens to be *v*). However, from the point of view of the operations themselves, neither "complementation" nor "specification" are of particular moment, both simply being instances of concatenation.

Let's make this point another way. Chomsky (1993, 1995a) distinguishes theta marking on the one hand from checking features like case, ϕ-features or Wh features on the other. Spec-X^0 is in the checking domain of a head, in contrast to the complement of X^0. However, if one does not invidiously distinguish theta assignment from case assignment or ϕ-feature or Wh-feature checking, all of these simply being instances of grammatical relations between elements, then there is no need to contrast checking from non-checking domains. What is relevant is the grammatical operation that licenses the interactions. The proposal here is that all such relations are discharged under concatenation. Most relations, with the general exception of the internal argument, will be established in Spec-X^0 configurations simply because theta domains are embedded within case domains which are in turn embedded within A′-domains. Why this is so is quite unclear (and it would be nice to know why we even have three separate kinds of domains (see Chapter 7)). But given that it is, there is no surprise that all except the internal argument is discharged in a kind of Spec-X^0 format. This follows from the requirement that grammatical relations be discharged under concatenation (coupled with labeling as construed above) and the assumption that Merge/Concatenate obeys binary branching. Thus, recent

[29] The empirical reasons for limiting agreement to Spec-X^0 configurations involve cases of agreement in the apparent absence of a Spec-X^0 configuration (as in existential constructions in English, or inverse agreement in Hindi and Icelandic). We discuss these cases and their implications for grammars incorporating AGREE versus Move in Chapter 6.

qualms about specifiers are correct in one sense; there is nothing special about them. But, on the present view, there is nothing grammatically special about complements either. Concatenation outputs both kinds of structures, with neither being more basic than the other except in the harmless sense that two applications of Concatenate creates more complex objects than one.

3.4 Why paths only include XPs

The conception of phrase structure proposed here which sees phrase building as the combined effort of concatenation and labeling understood in a bare phrase structure fashion has some interesting consequences when combined with a path theoretic understanding of minimality.

Chapter 2, recall, develops the following line of thought. Minimality as conventionally stated makes critical use of a c-command restriction as in (14).

(14) Minimality:
A movement operation cannot involve X^1 and X^3 over an X^2 which is relevantly identical to X^3 if X^2 c-commands X^3
$\ldots X^1 \ldots X^2 \ldots X^3 \ldots$

The restriction of minimality to c-commanding elements is derivable if one understands (14) in a broader context; namely, if minimality is construed as a principle that minimizes dependency lengths (as noted in Chomsky and Lasnik 1993: 89). The derivation makes two assumptions: First, paths are a good way to measure dependency length and second, that grammars are generally restricted to Boolean resources (specifically so when comparing paths). If we assume that the distance between two dependents in a sentence is measured by the union of the maximal projections that dominate each, then the restriction to Boolean resources for comparison will derive the c-command restriction on minimality mentioned in (13).[30]

A bonus of this way of construing minimality is that it effectively derives the A-over-A condition in terms of shortest dependency. (15) should make this point evident.

(15) $[_{X0P} \ldots X^0 \ldots [_{F1P} \ldots F2P \ldots] \ldots]$

Say that both F1P and F2P can interact with X^0 to check a relevant feature and assume that the grammar prizes short(est) dependencies measured as indicated above, then the path from F1P to X^0 must be shorter than the one from F2P to X^0 as the latter must include all the maxPs that dominate X^0 and F1P *plus* (at least) F1P itself. As this path contains the path of F1P as a proper subset, it is longer and so is blocked by the A-over-A. Thus, the A-over-A can be seen

[30] The reader is referred to Chapter 2 for details.

as just another instance of minimality once the latter is taken as minimizing dependency length.

How is all of this relevant here? It bears on our current divagations in two ways. First, the proposal in this chapter concerning labels explains why paths are measured in terms of *maximal* projections and second, it (in conjunction with some ancillary assumptions) provides a possible account of the Structure Preservation Condition (SPC). The SPC comprises three sub-conditions: (i) that maxPs target maxP positions (e.g. move from complement to Spec or Spec to Spec but never incorporate into heads), (ii) that intermediate X′-projections never move, and (iii) that heads only move to head positions (i.e. incorporate) and never target maxP positions (e.g. complement or specifier positions). Let's consider these two points in turn.

3.4.1 Computing shortest paths

Assume that paths are the way to measure distance within a phrase marker. The question still remains why we compute this distance in terms of maximal projections. Chapter 2 reviews some broad theoretical/empirical reasons for so calculating distance. However, there are other conceivable options. For example, why not say that a path consists of the union of *all* projections dominating target and launch site? The present conception provides a theoretical answer to this question. Consider a concrete case to help fix the problem.

(16) $[_{LP} \ldots L \ldots [_{BP}$ XP $[_{B'}$ YP [B . . .]]]]

In (16) the path from XP to L is the same as that from YP to L if we count only maximal projections, but the first path is shorter if we count all projections given that B′ dominates YP but not XP. So what one counts makes a difference and this is why it behooves us to have an answer to the question of why paths only include maxPs.

Note that the dilemma (should one count XPs only or all projections) dissolves once one takes a radically Bare Phrase Structure (BPS) conception of labels. BPS consists of two principal claims: (i) that bar-levels are relational constructs, i.e. no *intrinsic* grammatical difference obtains between various bar-level projections of a head; (ii) that grammatical computations only manipulate intrinsic (i.e. non-relational) features of grammatical terms.[31] In a word,

[31] BPS and the Inclusiveness Condition are tightly connected. Bar-levels are taken to violate inclusiveness as bar-levels cannot be lexical properties of an expression. Bar-levels can only be defined in the context of a phrase marker, a grammatical rather than a lexical construct. A question for the engaged reader: does labeling also violate inclusiveness? It is hard to say. The condition is amorphous. It requires that

> any structure formed by the computation . . . is constituted of elements already present in the lexical items selected for N [the numeration, NH]; no new objects are added in the course of the computation

grammatical operations only "see" inherent features and not relational properties. In effect, only the structure that BPS makes available is grammatically exploitable. In the current case, this implies that the labeling that the grammar sees is not the one in (16) but the one in (17).

(17) $[_{L} \ldots L \ldots [_{B}$ XP $[_{B}$ YP [B . . .]]]]

Note that the bar-levels have been wiped-out as this information is relational, not intrinsic given BPS reasoning. Let's now compute the paths between L and XP and YP assuming that it is the union of *all* nodes dominating target and launch site. The first path is {L, B}. The second looks initially different – {L, B, B} – but looks are deceiving as this is the same set as the set {L, B} (as sets never double count). In other words, counting *all* projections is identical to counting only maxPs once a radical BPS conception of labeling is adopted. The differences wash out once bar-level information is ignored, as is required under the labeling assumptions proposed here.

In short, the BPS approach to labels required to derive nesting (if Merge is simply Concatenate) also implies that paths can only be computed in one way or, much the same thing, any way of computing them leads to the same result. Or, BPS, which is necessary for the required interpretation of labels proposed here, yields the result that paths will only count one of any projection and this is why paths need only count XPs.[32]

3.4.2 *Minimality, BPS and structure preservation*

This conclusion leads to another question: does the grammar *ever* need to consider bar-level information to conduct its business? Given the reasoning above, the answer should be NO! If grammars cannot use relational properties of phrases to compute paths, they should be blind to relational properties

> apart from rearrangements of lexical properties (in particular, no indices, *bar levels in the sense of X-bar theory* [my emphasis, NH] . . . this condition holds (virtually) of the computation from N → LF . . . (Chomsky, 1995a: 228)

Is labeling as understood here in conformity with these desiderata? Well, there are no bar levels, as we insisted above. Note too that labels just *are* lexical items and labeling is construed as mapping complex concatenates to one of the concatenates involved (viz. the head) so as to be able to further concatenate the complex expression. In fact, as noted earlier, one way of looking at the labeling proposed here is that it serves to close the operation of concatenation in the domain of the lexical items. Just as addition is closed in the domain of the natural numbers (i.e. adding any two natural numbers returns another natural number), so too concatenating *and labeling* any two lexical items returns an expression type identical to a lexical item (and hence allows further concatenation). If this is correct, then there is a sense in which the current proposal obeys Inclusiveness. Of course, there are surely other interpretations of Inclusiveness where the labeling proposed here is excluded. I leave the evaluation of these matters to the exegetically inclined. For relevant discussion, see Seely (2006).

32 Of course, given this logic, counting heads would have worked out just as well as phrases have one head and one XP projection.

in general. That means that the grammar should never advert to maxPs. On the face of it, this seems clearly false. For example, the Structure Preserving Condition (SPC) limits maxP movement to maxP positions and limits the X^0 movement to X^0 positions. Similarly, it has generally been assumed that intermediate X's are not themselves subject to movement operations. These conditions all seem to advert to bar-levels and hence to relational properties (if BPS is correct).

However, consider what happens if we combine BPS with the idea that grammars minimize dependency length (measured by path length as above) and continue to assume that all grammatical interactions involve the intrinsic features of expressions, i.e. that no relational properties are relevant and that labeling is an operation according to which labels are just lexical items that name a concatenate (thus turning complexes into atoms of the kind named by the head). For the time being let's also set head movement to one side. With this in mind, consider (17) repeated here as (18).

(18) $[_{L}$ . . . L . . . $[_{B}$ XP $[_{B}$ YP [B . . .]]]]

Say that L and B have features that need checking by inter-relating them. They are inter-related by copying B and concatenating it with L. Which B atom? Recall that we must minimize dependency length. So "Which B?" translates here as the question of which projection of the B atoms in (18) is to be copied. There are three possibilities: the simplex atom B, the complex atom labeled B containing YP, and the complex atom labeled B and containing XP and YP. If grammars minimize dependency length, the one that is copied must have the shortest path to L. This must be the last of the three mentioned alternatives. Why? Because its path is clearly the shortest as it does *not* contain B. The paths of the other two contain B as B dominates each. Thus, given our assumptions, the largest B atom will be copied and concatenated with L creating (19).

(19) $[_{L}$ $[_{B}$ XP $[_{B}$ YP [B . . .]]] $[_{L}$ . . . L . . . $[_{B}$ XP $[_{B}$ YP [B . . .]]]]]

Observe that this amounts to requiring that XP move but without having to mention that it is a maxP. Rather that this is the B-constituent that moves falls out from more general assumptions concerning labeling and the minimization of dependency length.

Note too that the reasoning outlined here implies that intermediate X′s should *never* move. Why so? Because moving an intermediate projection will always violate the A-over-A in the sense of having a longer path than would ensue were the "maxP" moved. Once again this only works if we assume that labels are identical to heads. In (19) it is critical that the projections of B have the very same properties that B itself has. This follows if labels are complex atoms with properties identical to the heads that label them. In other words, we are here taking quite literally the X′-theoretic intuition that phrases are projections of heads and that labels are heads.[33]

In sum, our assumptions seem to derive two parts of the SPC: that XPs move to XP positions and that X′s don't move. We need a further assumption to block incorporation of maxPs into heads. To block this we can adopt a version of Chomsky's morphological condition (1995a: 319): morphology only deals with lexical heads and their features. Chomsky understands lexical heads to be X^0s. We will understand them to be simplex lexical atoms (i.e. LIs (atoms) in the numeration or in the lexicon) and congeries of simplex atoms. The latter is required to allow successive head movement. What is critical here is that the morphology places this condition on the grammar. It is a morphological bare output condition. It suffices to block incorporation if maxPs into Y^0s as maxPs (and X's) are too complex for the morphology to handle.[34]

A further consequence of the current proposal, taken as stated, is to rule out X^0 movement. As with the case of X′ movement, the path from X^0 to the target will always be longer than the path of the XP to the target.[35] It is not clear if this is a desirable consequence or not. It has been proposed that head movement is not a part of the core computational system but is rather a kind of PF process.[36] If this suggestion proves to be correct, then the fact that head movement violates minimality would be a welcome result for it would explain *why* head movement was not a possible grammatical operation. In

[33] That labels are heads is suggested in the discussion in Chomsky (1995a: 244–245).

[34] Morphological incorporation is almost certainly more subtle than the discussion here suggests. For relevant discussion, see Nunes (2004) and Boeckx, Hornstein and Nunes (2007). For current purposes, it suffices that we can incorporate an analogue of Chomsky's morphological principle suitably refined to be empirically adequate.

[35] Željko Bošković (p.c.) also points out that this is incompatible with the injunction to move the slightest amount of material to meet a grammatical requirement (which we have dubbed the "Lightest Load Principle" (LLP)). Whether the LLP holds for overt syntactic movement operations is unclear. The bulk of the arguments in its favor pertain to LF movement. However, it is worth observing that on the conception here it is always "atoms" (simplex or derived) so perhaps the LLP, if it exists, is vacuously satisfied in overt syntax.

[36] See Chomsky (2000) and Boeckx and Stjepanović (2001). There are various ways to implement this position. One could assume that head "movement" is actually the morphological reflex of head-to-head selection/agreement. This is similar to the Conflation operation in Hale and Keyser (2002) and it generalizes the suggestion in Lightfoot (1991) and Bobaljik (1995a) for affix hopping languages like English and extends it to head movement languages like French. The idea is roughly as follows. In both English and French finite T^0 and *v*P merge and selection/agreement is established. In English the morphological reflex of this operation is realized on the *v*+V structure while in French it is realized on the T^0. On this conception, head movement is a morphological operation and so outside the confines of the grammar properly speaking. This is consistent with the claim that head movement cannot apply in the grammar as it would violate the A-over-A condition. See Harley (2004) and Boeckx (2006) for a fuller discussion.

effect, the absence of head movement would constitute additional support for the proposals above.[37]

What, however, if head movement is part of the grammar? Does this argue against the conception outlined above? Not necessarily. Consider the following dodge. Say that the target of movement is a simplex atomic head, X^0 and assume that a version of Chomsky's morphological condition (viz. that only simplex atoms can incorporate into simplex atoms) is correct. Under these conditions a complex atom cannot successfully move for it is morphologically prohibited from incorporating into the X^0-target. Only the simplex head will be able to licitly combine with X^0. Under these conditions, the A-over-A reasoning that forced movement of the maxP is (plausibly) inert as the complex atom cannot do what the simplex one can.

Observe that this reasoning implies there can be only one kind of X^0 movement; one in which X^0 moves to a Y^0 into which it incorporates. Thus, heads can *only* move in a head-to-head manner for only in this case can they do what the larger atomic complex cannot and thereby evade the A-over-A condition.

There is one other way of allowing head movement compatible with the assumptions here. One could adopt a Kayne like conception of head movement as actually a species of maxP remnant movement with all but the head removed from the maxP. As this assimilates X^0 movement to maxP movement it fits easily within the current set of assumptions.

Thus, there are various ways to permit head movement compatible with the present set of ideas and thus to derive the three parts of the structure preservation constraint without adverting to bar-levels. In sum: assuming the A-over-A and a BPS conception of labels, we derive that XPs always move (as they traverse the shortest paths) unless the target of movement is a head, in which case only the head can move due to Chomsky's morphological restrictions (which prohibits complex atoms from incorporating). Thus, structure preservation is derived: only heads move to heads (if they move at all), XPs move everywhere else and X′s never move at all.

If this is correct, it eliminates the need for the Chain Uniformity Condition (CUC).[38] This, I believe, is a welcome result. The CUC is an odd condition on several counts. First, it adverts specifically to chains as computational objects and there are problems with so conceiving them as, for example, they violate the Inclusiveness Condition (among other problems).[39] Second, the CUC overtly

[37] Head movement could then be treated as an instance of morphological agreement along the lines of Lightfoot (1990) or Bobaljik (1995b). This would suffice for standard cases of local A-movement. It is less clear that this could extend to cover cases of long head movement of the kind found in Vata (Koopman 1984, Nunes 2004). Nunes (2004) argues that here the V moves to an FP position where it must incorporate into F^0. This would then involve a kind of non-local agreement that is less amenable to a morphological analysis.

[38] See Chomsky (1995a). The CUC is intended to cover the empirical effects of the SPC.

[39] For discussion, see Epstein and Seely (2006), Hornstein (2001), and Lasnik (1999).

refers to bar-levels (it states that chain links must be uniform with respect to bar-level) and this goes against the spirit of BPS especially as interpreted above, viz. as eschewing the use of relational information within the computational system. Third, it essentially stipulates what one wishes to derive; namely that XPs move only to XP position and X^0s to X^0 positions.[40] Thus, the CUC is worthy of elimination.

Last point: note that the dodge that allows an X^0 to move will not license the movement of an intermediate X′. It cannot move under the imagined conditions as it is not simplex. Thus, X′s should never be targets of movement given the above reasoning.

One objection to the proposed dodge is that it requires distinguishing complex and simplex elements and this might seem to be simply reintroducing bar levels in a new guise. However, this strikes me as incorrect. Recall that the projections of a head are atoms for the purposes of concatenation; they are clearly not atomic with respect to other operations (e.g. Copy) or (possibly) with respect to the interfaces. Nothing proposed here implies that complex elements are not complex from the purview of the CI or AP systems. Thus, if the condition limiting incorporation to heads is a morphological condition, then nothing proposed here prevents the morphological system from distinguishing complex atoms from simplex ones. How morphological failure is used to compute potential movements is a delicate matter, one that I will put aside for now. For current purposes, it suffices that simplex atoms are always morphological simples whereas X′s and maxPs are typically not.

In sum, it seems that there is some room for maneuver should head movement prove to be a resilient grammatical operation. This said, it would also be a very congenial result should X^0 movement prove to be a PF or morphological effect and not part of the syntactic computational repertoire.

3.4.3 An interpretive interlude

It is perhaps worth observing that the system outlined here can be taken as an elaboration of some of Chomsky's current ideas. He has conceived of displacement as triggered by features on heads with phrases being carried along in a pied-piping manner by the moving head. It has, however, been somewhat unclear what might be meant by "pied-piping" here and why it is that what is carried along for the ride is co-extensive with the maxP of the head. The story

[40] That XPs cannot adjoin to Y^0s is consistent with uniformity. This is ruled out by the morphological condition (which, recall, I adopt as well). In Chomsky (1995a) head-to-head incorporation violates the CUC as the incorporated head will be interpreted as maximal given the relational definitions in BPS. Chomsky (1995a) circumvents this apparent problem by postulating a second morphological condition that prevents CUC from applying inside words, i.e. incorporation structures.

we have sketched above mechanizes Chomsky's suggestion. On the present account, the information carried by a head is literally carried at all levels of projection. Thus, maxPs have all the information of the head that it is "projected" from (i.e. the head that labels the complex). Moreover, the reason the maxP moves is for minimality reasons and not because the maxP itself is manipulated by the grammar. In other words, the present framework is compatible with the idea that only heads enter into grammatical relations and that phrases move as by-products of head-to-head interactions.[41]

A second point is also noteworthy. It was important to the above discussion to ignore relational properties in computing grammatical moves. The obvious suggestion to explore is that this holds not only in the case of phrase structure relations (e.g. bar-levels) but for *all* grammatical relations. In other words, a strong condition on potential grammatical operations is that it be limited to the inherent properties of the lexical inputs (this is consonant with a strong reading of the Inclusiveness Condition). This would effectively bar grammatical operations that converted objects into subjects or indirect objects into objects (as in relational grammar and LFG) as notions like "subject," "object," and "indirect object" are relational. This could plausibly also forbid operations defined over thematic elements like agents, themes, goals, etc. as these are also plausibly relational and even nominatives, accusatives, datives, etc. for the same reason.[42] Grammatical operations might concatenate Ds and Ns or Vs and Ds or Cs and Ts as these involve inherent categorical information of lexical items. Features like +/− human or +/− animate might also be grammatically visible as would be the standard phi-features. However, pure relational features would be barred as violations of a very strong version of the Inclusiveness Condition. Relational notions are not intrinsic features of lexical items, so if grammars can only manipulate lexical features, these cannot be manipulated. If some version of this idea can be maintained then grammars, though producing representational objects for interpretation at the interfaces, would nonetheless not exploit properties of representations for grammatical ends. In this sense, they would be best understood in derivational terms.[43] Whether it is possible

41 The analogy to pied-piping is actually misleading. Pied-piping occurs when the phrase that moves is featurally distinct from the phrase that is the motive force of the movement, e.g. when a PP moves though it was a Wh DP that has the relevant features. This is not subject to a minimality analysis in any straightforward way, in contrast to the cases discussed in the text.

42 This does *not* imply that notions like "subject," "object," "agent," "theme," etc. are linguistically unimportant. Even if the grammar were to religiously ignore relational information, nothing said here (even if true) would prevent the interfaces from interpreting the structures that the grammar created using relational notions. In fact, if the CI system exploits an eventish logical form (cf. Pietroski 2005) then given that events participants are understood in grammatical function/thematic terms, we would hope that the objects created by the syntax would allow for the identification of subjects, objects, etc. See Chapter 4 for discussion.

43 Chains would also not be grammatical objects on this view as they are inherently relational (see Hornstein 1999, 2001 for some discussion about chains violating inclusiveness as well as

for grammars to entirely dispense with relational notions is, I believe, quite unclear. The central question, I believe, will revolve around how to understand "Least Effort" and "Greed." How do features function to drive grammatical operations? Which features are relevant? How do the interfaces interact with grammatical operations? These are very big questions and much beyond the scope of the present chapter (and book).[44]

3.4.4 *Phrases have only a single head*

It has long been assumed that phrases cannot be multiply headed. Chomsky (1995a) attempts to derive the fact that phrases have unique heads from the idea that labels are just feature sets and that the standard Boolean combinations of features that would result if two heads donated the label would be incoherent. Thus, if one allowed a label to be the intersection or union of the feature sets of the heads many times, one would either get incoherence in the label or nothing in the label: when, for example, the features negated each other – say V is $\{+V, -N\}$ and N is $\{-V, +N\}$ – then the union might be seen to be incoherent as it is $\{+V, -V, +N, -N\}$ and the intersection of the two sets would be null.[45] Even granting this point (but see note 44), whether the reasoning holds for the general case is unclear. Do the features of a functional head clash with those of a lexical complement? If not, can we find multi-headed structures in these cases? If not, then this ingenious proposal won't work. Interestingly, the approach developed here prohibits multi-labeled (i.e. multi-headed) phrases if, as proposed, labeling returns a concatenative atom. A double-headed element

Epstein and Seely 2006 and Lasnik 1999). Conditions like Chain Uniformity would be doubly cursed given their being defined over chains *and* adverting to uniformity of *bar*-levels, another relational notion given BPS. This further motivates the reanalysis of CUC as above.

44 Nonetheless, it is worth noting that the status of Greed is far less clear in current discussions than it was in the theories of the mid 90s. Greed is a way of implementing the Last Resort notion, viz. the idea that operations apply because they have to meet interface requirements. However, it goes beyond Last Resort in breaking a global interface requirement down into smaller local step by step feature checking requirements. This allows derivations to move from a generate-and-filter format to a build-structure-only-if-licensed format. Features then are the mechanisms through which "global" interpretive effects of the interface are used to "locally" license structure building. Seen in this way, the question of what features exist amounts to the question of what kinds of information grammars have developed to track structure building. For a further brief discussion of this in the context of ATB operations, see Hornstein and Nunes (2002). It would not be surprising to find that a host of interpretive relations that typically "emerge" at the interface would be codified in a feature scheme that the grammar would locally exploit.

45 Actually, the apparent argument is not that clear here. The sets of features are not *in themselves* incoherent. After all, the set $\{1, -1\}$ is perfectly fine. If so, more needs to be said about what makes the set $\{+V, -V\}$ incoherent. The argument presumes some interpretation of these sets of features so that a given expression cannot have both +V like properties and –V like properties. To flesh out the argument requires stating how these feature sets are interpreted so that the incoherence becomes manifest.

would be read as a non-lexical hybrid AB. For example, (20) says that A^B is an AB. On the assumption that ABs are not themselves atoms (note As are and Bs are but ABs are not) then having two heads label a phrase prevents that phrase from being input to further concatenative operations. Thus, multiple labels (and so multiple heads) are generally ruled out.[46]

(20) $[_{AB}$ A^B]

Whether this is a good result will depend on whether multi-headed structures exist. It is often assumed that this is impossible (Chomsky 1995a, Kayne 1994). Chomsky's (1995a) proposal might leave room for this (it will depend on how the features of heads combine). The current proposal leaves virtually none.[47]

3.4.5 *Why grammatical operations target constituents*

One of the signature properties of natural language is that grammatical rules target constituents. The account outlined above that treats Merge as a species of Concatenate and Move as the composite of Copy and Merge accounts for why this phrase structure and movement manipulate constituents. Let's see why.

If Merge is a species of concatenation and concatenation is restricted to primitive lexical items or derived ones, viz. labeled concatenates, then Merge can only manipulate constituents, i.e. lexical items or labeled concatenates. These are the only grammatical "terms" in the sense of Chomsky (1995a) as only these expressions can be concatenated.

The present proposal also limits movement to concatenables. Movement, recall, is just the combination of Copy and Merge (= Concatenate). Concatenate is only defined for atoms. Thus, for a copy to be integrated into a structure, it must be a concatenative atom. As concatenative atoms are constituents, it follows that only constituents move.[48]

This reasoning can be extended to rules of construal like obligatory control and reflexivization if these are also products of movement as suggested in Chapter 2. The antecedent of an obligatory controlled PRO or a local

[46] It should be permitted in structures that need not concatenate further. However, on the assumption that derivations only converge if single rooted, this seems to leave only the whole derived sentence as potentially multi-headed. It is not clear to me how this prediction could be tested as the root clause is not subject to the sorts of operations that we use to determine a phrase's head, e.g. selection or subcategoriztion.

[47] Though multi-headed structures are barred, it may be that dual structures are not. By "dual structures" I mean the sorts of labeling discussed in Hornstein and Uriagereka (2002) as re-projections. Might one and the same phrase have dual structures and "resonate" between them? If one could make sense of this idea, then something with the empirical properties of multi-headed structures might be theoretically possible even though multi-headedness *per se* was barred. These inchoate conceptions await further research.

[48] As noted above, this allows the Copy operation to be maximally general: Copy anything, the need to concatenate will assure that only constituents get copied.

reflexive is typically a single DP. Split antecedents are generally excluded (**Everyone*$_1$ *asked someone*$_2$ *about themselves*$_{1+2}$). This follows if an obligatory control PRO of a local reflexive is the residue of the movement of its antecedent.[49] Interestingly, if pronoun binding is *not* a product of movement (e.g. contra Kayne 2002, cf. Chapter 1), then they should be bindable by non-constituents, as seems to be the case (*Everyone*$_1$ *told someone*$_2$ *that they*$_{1+2}$ *should wash themselves*$_{1+2}$). But this observation must be treated with caution. Non-constituent antecedents should also be possible if the bindee is complex. Thus, if plural pronouns are actually complex conjuncts of sorts (e.g. doubled expressions) then each conjunct could have a unique antecedent, though it would appear that the plural had binders that were not constituents. This would be consistent with a movement-based account of pronominal binding.[50]

We can extend this reasoning to ellipsis, if it also involves movement, at least in some cases. Johnson (2001) suggests that VP ellipsis involves movement to Topic with subsequent deletion of the moved Topic. This accounts for the parallel between examples like those in (21) and (22).[51]

(21) a. He would eat pizza and Holly would too
 b. *He would eat pizza and Holly ~~would eat pizza~~ too

(22) a. Eat pizza, Holly would
 b. *would eat pizza Holly

If this is correct, then we would expect VP ellipsis to target constituents because it involves movement which only targets constituents (due to its having Concatenate as a subpart and due to Concatenate only being applicable to concatenative atoms).

This reasoning suggests that other grammatical operations, ones that do *not* involve concatenation, need not be limited to constituents.[52] There is some evidence that some instances of sluicing need not target constituents. Yoshida (2006), for example, notes that sluicing is possible in cases like (23).

(23) John kissed someone without knowing who

This seems like a fine sluice but it is not clear how it can be given that what is sluiced is a TP minus the adjunct *without* clause. If this hangs within TP, as generally assumed, it appears that a non-constituent has been sluiced. Curiously, TP cannot be topicalized (*This book, I said that John read vs.* **John read this*

[49] There are complications that I am abstracting away from. See Boeckx, Hornstein, and Nunes (in progress) for a review of split and partial control.
[50] This is proposed in Kayne (2002) and Vassilieva and Larson (2001).
[51] For further examples and discussion see Johnson (2001). See also Donati (2003), Fitzpatrick (2006), Kayne (2005), Rizzi (2005), and Szczegielniak (2005).
[52] This consequence was brought to my attention by Howard Lasnik (p.c.).

book, I said that) and this suggests that Topicalization of the elided TP does not feed sluicing.

Similar apparent deletions of non-constituents obtain in the case of antecedent contained deletion (ACD). Consider two examples. In examples like (24) it appears that the "antecedent" of the elided constituent is not itself a PF constituent.

(24) John expects everyone that I do to be at the reception

Here the "antecedent" for the elision is *expect to be at the reception* ((24) means *John expects everyone that I expect to be at the reception to be at the reception*). But this is not a surface constituent. If this form of VP ellipsis involves deletion under some form of syntactic identity, then it appears that whatever operation underlies this deletion is not restricted to constituents.[53] In fact, this seems to be generally true of ACD ellipsis as none of the elided elements in (25) are generally assumed to be identical to *surface* constituents.

(25) a. John blamed everything that I did on Sam
b. John talks about whoever I do

In (25a) what has been elided is *blamed on Sam* and in (22b) it is *talk about*. Neither is a surface constituent.[54] It is noteworthy that VP topicalization and relative clauses do not mix well. This is consistent with the view that when movement feeds ellipsis it targets constituents but otherwise it need not.

(26) a. John saw me everytime that I kissed Mary
b. *Kiss Mary John saw me everytime that I did
c. *John saw me everytime that kiss Mary I did
d. John likes the place where I kissed Mary
e. *Kiss Mary John likes the place that I did
f. *John likes the place where kiss Mary I did

The above is not intended to argue for the grammatical manipulation of non-constituents so much as observe that whereas movement (plus grammatical operations plausibly fed by movement) seems to respect constituency, it is less clear that the operations underlying some forms of ellipsis similarly do. The present account is consistent with distinguishing among grammatical operations. If constituency is tied to concatenability, then grammatical manipulations that do not involve this operation need not target constituents.

[53] See Merchant (2001) and Lasnik (2001) for arguments that ellipsis is a deletion operation.

[54] This does *not* mean that some LF constituent might not be involved in licensing ellipsis. However, *if* ellipsis involves deletion of phonetic material (rather interpreting null structure at LF) then part of ellipsis involves the apparent deletion of phonetic material that are not constituents.

3.5 Conclusion

This chapter has suggested that labeling (understood in BPS terms) when combined with Concatenate and Copy, results in systems of rules that have the broad qualitative outlines of natural language grammars. In particular concatenation, labeling and minimizing dependency length conspire to yield grammars that display unbounded nested recursion, the endocentricity condition, movement, uniquely headed phrases, structure preservation and a concern with constituency. Such grammars have a host of other attractive properties as well.[55] We consider them all together in Chapter 5 in the context of the following question posed in Hauser, Chomsky and Fitch (2002): what features of the faculty of language (FL) are unique and which are common across the cognitive domains?

55 Another feature of the present system is that the derivations are monotonic. This follows from the fact that they enforce a strong version of the Extension Condition. Monotonicity is generally taken to be a pleasant feature of computational systems. A nice question, one that I have not pursued here, is the degree to which computationally attractive properties fall out of the kind of system outlined above. Minimizing dependency length and monotonicity are nice features for a computational system to have. Both are central to the system proposed here.

4 Some thoughts on adjunction[1]

4.1 Introduction

It is fair to say that what adjuncts are and how they function grammatically is not well understood. The current wisdom comes in two parts: a description of some of the salient properties of adjuncts (they are optional, not generally selected, often display island (CED) effects, etc.) and a technology to code their presence (Chomsky-adjunction, different labels, etc). Within the Minimalist Program (MP), adjuncts have largely been treated as afterthoughts and this becomes evident when the technology deployed to accommodate them is carefully (or even cursorily) considered.

The primary aim of this chapter is to propose a phrase structure for adjunction that is compatible with the precepts of Bare Phrase Structure (BPS). Current accounts, I believe, are at odds with the central vision of BPS and current practice leans more to descriptive eclecticism than to theoretical insight. I have a diagnosis for this conceptual disarray. It stems from a deeply held though seldom formulated intuition; the tacit view that adjuncts are the abnormal case while arguments describe the grammatical norm. I suspect that this has it exactly backwards. In actuality, adjuncts are so well behaved that they require virtually no grammatical support to function properly. Arguments, in contrast, are refractory and require grammatical aid to allow them to make any propositional contribution. This last remark should come as no surprise to those with neo-Davidsonian semantic sympathies. Connoisseurs of this art form are well versed in the important role that grammatical (aka, thematic) roles play in turning arguments into modifiers of events.[2] Such fulcra are not required for meaningfully integrating adjuncts into sentences. In what follows, I take this difference to be of the greatest significance and ask what this might imply for the phrase structure of adjunction.

[1] This chapter is based on joint work with Jairo Nunes. A version of the material contained here appears in Hornstein and Nunes (2008).

[2] See Higginbotham (1986), Parsons (1990), Pietroski (2005), and Schein (1993) for extensive discussion.

A second boundary condition in what follows is that an adequate theory of adjunction comport with the core tenets of BPS. Current approaches sin against BPS in requiring an intrinsic use of bar-levels and in using idiosyncratic labeling conventions whose import is murky at best. We rehearse these objections in the following sections. A goal of a successful theory of adjuncts should be to come up with a coherent account of adjunction structures that (at least) allows for a relational view of bar levels along the lines of Chomsky (1995a) (following earlier suggestions of Muysken 1982).

More ambitiously, one could require that the bar-level properties of adjunction structures play no grammatically significant role. Chapter 3 proposed a very strong version of the Inclusiveness Condition, one in which *only* intrinsic features of lexical elements are used by the computational system. This excludes, among other things, bar-level information (which is relational) from the purview of the syntax.[3] Thus syntactic rules cannot be stated in terms like "Move/delete XP" or "Move X^0" or "never move X′," etc. Relational information *may* be important, at the interpretive interfaces for example, but syntactic computations *per se* cannot exploit these relational notions (given a strong version of the Inclusiveness Condition), as they are not intrinsic features of lexical items. In what follows, we adhere to this strong version of the Inclusiveness Condition.[4]

The chapter is organized as follows. Section 4.2 reviews the general properties of adjunction structures assumed in the literature and argues that their standard account in terms of Chomsky-adjunction is not easily accommodated within the BPS approach to adjunction in terms of a distinct labeling procedure. Section 4.3 discusses what goes wrong if adjunction structures are assigned the same label as non-adjunction structures and Section 4.4 argues that the output of a Merge operation need not be labeled and this is crucial for the distinction between arguments and adjuncts. Section 4.5 discusses some consequences of this proposal and Section 4.6 offers a brief conclusion.

4.2 General properties of adjunction structures

Prior to minimalism, adjunction was an operation that returned a phrase of the same type as the one the operation had targeted. (1) formally illustrates (Chomsky-) adjunction with respect to phrases.

(1) $[_{XP}\ [_{XP}\ [_{XP} \ldots X^0 \ldots]$ adjunct] adjunct]

(2) $[_{VP}\ [_{VP}\ [_{VP}$ read a book] quickly] in the yard]

(3) $[_{NP}\ [_{NP}$ student of physics] from France]

[3] See Chapter 3 for discussion.

[4] This version of the Inclusiveness Condition suggests a strong reading of the autonomy of syntax thesis. If correct, syntactic operations are blind to certain kinds of information that the interfaces may exploit. This makes the divide between syntax and the other components of the faculty of language rather broad.

(2) and (3) exemplify the structure in (1) with the adjuncts *quickly/in the yard* and *from France* adjoining to VP and NP, respectively, and returning VP and NP, respectively. Accounts differed on whether adjuncts adjoined to XPs or to X′s. However, they agreed in assuming that the output of adjunction left the input labeling (and constituency) intact.

The labeling in (1)–(3) codes five important properties criterial of adjunction. First, adjunction conserves bar-level information. Note that in (1)-(3) adjunction leaves the maximality of the input VP intact and in this regard, it contrasts with complementation as the latter changes bar-level information. For example, in (2) a V^0 *read* combines with a NP *a book* to yield a VP (not a V^0). Second, adjunction leaves the category information intact. If the input is verbal, the output is verbal. Third, headedness is preserved. Thus, the head of the complex in (1) is X^0, the head of (2) is *read*, and the head of (3) is *student*. Forth, the adjunction structure "inherits" the bar-level information of the target. Thus, in (2), we have three maxV projections: *read a book, read a book quickly* and *read a book quickly in the yard*. Last of all, there is no apparent upper bound on the number of adjuncts. Once again this contrasts with arguments where there is generally an upper bound of three.

These five properties are well grounded empirically. The preservation of categoricity and headedness tracks the fact that adjoined structures do not introduce novel subcategorization or distribution relations. For example, in (4a) below perfective *have* selects/subcategorizes for a perfective *–en* marked V. This selection requirement is unchanged in (4b) despite the adjuncts.

(4) a. has/*is [$_{VP}$ eaten a bagel]
  b. has/*is [$_{VP}$ [$_{VP}$ [$_{VP}$ eaten a bagel] quickly] in the yard]

On the standard assumption that only heads can be seen by elements outside an XP and that heads mark the category of a complex phrase, the data in (4) indicate that the complex complement of *has* in (4b) is a VP projection of the perfective head *eaten* (as in (4a)). The same argument can be made in the nominal domain. For example, (5a) shows that *these* demands a plural nominal head and (5b) shows that adding nominal adjuncts does not change this requirement.

(5) a. These [$_{NP}$ students/*student of physics]
  b. These [$_{NP}$ [$_{NP}$ students/*student of physics] from France]

Nor does adjunction affect the distribution of expressions. Thus, if an XP can occur in some position, an XP modified by any number of adjuncts can, as well. For example, predicative NPs can occur in (6a) and the more complex NPs in (6b) can, too.

(6) a. John is a student of physics
b. John is a student of physics from France

The conservation of bar-level reflects a different set of facts, two kinds actually. If an XP can be target of a grammatical operation (e.g. movement, ellipsis, or anaphoric dependency), then adjunction does not remove this property. Thus, VP fronting can apply to the VP *eat the cake* in (7a) and can still apply to it in (7b).[5]

(7) a. John could [eat the cake] and [eat the cake] he did
b. John could [eat the cake] in the yard and [eat the cake] he did in the yard

Thus, the VP status of *eat the cake* is not disturbed by adjoining *in the yard* to it. In addition, the VP plus adjuncts are also VPs as they too can be fronted.

(8) a. . . . and eat the cake in the yard he did with a fork
b. . . . and eat the cake in the yard with a fork he did

Similar effects are attested with VP ellipsis, *do-so* anaphora, and *one* substitution, as shown in (9) and (10) below. These each target the head+complement (obligatory) plus any number of adjuncts (optional).

(9) John ate a cake in the yard with a fork and
a. Bill did (so) too
b. *Bill did (so) an apple in the hall with a spoon
c. Bill did (so) in the hall
d. Bill did (so) with a spoon
e. Bill did (so) in the hall with a spoon

(10) This [[[student of physics] with long hair] from France] and
a. that one
b. *that one of chemistry (with long hair from France)
c. that one from Belgium
d. that one with short hair
e. that one from Belgium with short hair

The fact that the complement cannot be left out in (9b) and (10b) is attributed to the fact that the head sans complement is not an XP and so not a target of the relevant operation. The fact that any number of adjuncts can optionally be targeted follows if head and complement plus any number of adjuncts are all XPs and thus of the same size (measured in bar-levels).

To recap, the classical approach to adjunction captures several salient properties: it preserves the bar-level information of the target, retains the category

[5] See 4.4 below for some discussion on head-to-head adjunction.

information and headedness of the target in the adjoined structure, returns a constituent with a category label identical to that of the target, and can do this without limit. The labeling convention in (1) succinctly summarizes these facts by having adjunction label the output of the adjunction operation with the same label as the target/input.

From an MP perspective, this standard account of adjunction structures is unsatisfactory because it is incompatible with BPS views concerning bar levels and the Inclusiveness Condition. To see this, consider the fact that adjunction leaves the maximality of the target XP intact. In BPS, a projection is maximal if it no longer projects. However, the conservation of headedness in adjunction structures implies that the head of the input is also the head of the output. But this is incompatible with BPS if we also insist that the XP that projects still retains its XP status. Thus, from a strict BPS perspective, either head properties are not conserved in adjunction structures or the XP to which the adjunct has adjoined becomes nonmaximal after adjunction. Similar considerations apply to XPs associated with multiple adjunctions. Take (1), repeated below in (11), for instance. Given a BPS understanding of bar-levels as relational, only the outmost XP can be maximal; crucially, the "intermediate" adjoined projection cannot be maximal if conservation of headedness is preserved in the larger structure.

(11) $[_{XP} [_{XP} [_{XP} \ldots X^0 \ldots]$ adjunct] adjunct]

This would seem to present BPS with empirical problems for we noted above that there is interesting empirical evidence that each of the XPs in (11) can function as targets of the same operations. We also found evidence that the selection properties of (11) are identical to those of the simple non-adjoined XP in (12).

(12) $[_{XP} \ldots X^0 \ldots]$

This suggests that the head of (12) is the same as that of (11). There is, thus, a *prima facie* incompatibility between BPS, the classical approach to adjunction in terms of Chomsky-adjunction, and the facts.

MP has a different account of adjuncts. It proposes that adjuncts are labeled differently from complements.[6] As Chomsky (1995a: 248) puts it:

[6] In fact, Chomsky's (2000) distinction between set-merge (for arguments) and pair-merge (for adjuncts) suggests that not only the output of the merger operation may be different depending on whether we are dealing with an argument or an adjunct, but the merger operations themselves may be of a different nature. From a methodological point of view, the best situation would be that there is nothing that distinguishes the operation that merges arguments from the one that merges adjuncts. Another possible interpretation of Chomsky is that the interpretive result of pair-merge is an ordered pair while that of set-merge is an unordered set. Thus, the operation of pairing the arguments and adjuncts may not differ though the output may, say due to labeling. This option is discussed in Hunter (2008). See 4.4 below for further discussion.

Substitution forms L = {H(K), {α,K}}, where H(K) is the head (= the label) of the projected element K. But adjunction forms a different object. In this case L is a two-segment category, not a new category. Therefore, there must be an object constructed from K but with a label distinct from its head H(K). One minimal choice is the ordered pair <H(K), H(K)>. We thus take L = {<H(K), H(K)>, {α,K}}. Note that <H(K), H(K)>, the label of L, (. . .) is not *identical* [NC's emphasis; NH] to the head of K, as before, though it is constructed from it in a trivial way.

Given this notation, an adjunction structure would look like (13):

(13) $[_{\langle X, X\rangle}\ [_{\langle X, X\rangle}\ [_{X(P)} \ldots X^0 \ldots]\ \text{adjunct}]\ \text{adjunct}]$

The passage above discusses segments versus categories, a distinction that we will return to anon. For now observe that the label of an adjoined structure is *different* from that of the element that it is adjoined to. Thus the *head* of the adjunction structure is distinct from that of the constituent adjoined to. If one takes this to mean that the head of the target of adjunction has *not* projected, then one of the problems noted above for the classical theory can be addressed.[7] As the labels differ (i.e. the heads did not project), given BPS the inner X(P) and the outer <X, X> categories are both maximal, thus being compatible with the movements in (7b) and (8b). However, this result is achieved at a price of redundancy, as VP movement now resolves into two different operations – <X, X> movement and X(P) movement – at least if operations are distinguished by the objects they apply to.

Moreover, the <X, X> notation still leaves several unresolved questions. For example: what is the status of the inner <X, X> projection in (13)? Is it maximal or not? If it is, then why does it determine the label of the outer projection? On the other hand, if it is not maximal, we would expect it to function differently from the outer projection, but so far as we can test this, the two function identically. Thus, given that the outer adjunction projection in (8b), for instance, can move, so can the inner one, as shown in (8a). More generally, if the labels of adjunction structures differ from those of their targets, then how do we account for the fact that their distributional properties are identical? Why are they subject to the same selectional restrictions? Why do they behave alike with respect to grammatical rules like ellipsis, movement, or anaphora? To put this same point more baldly: if the labels of adjunction structures are not *identical* to the labels of the non-adjunction categories that they target, why is it that the properties of the two kinds of constituents are indistinguishable?

The issues reviewed here show that the BPS approach to adjuncts in terms of distinct labels misses the generalizations that the classical theory coded. The

[7] Whether the head has projected is actually unclear given Chomsky's observation that the label of the adjunct is constructed from the head of the adjoined-to in a "trivial" way. Still, given Chomsky's underscoring of the fact that the two labels are distinct (not *identical*), it appears that he would not see the label of the adjunction structure as the same as that of the adjoined-to.

trouble seems to be that the labeling that has been proposed relies on bar-level information in a crucial way. But this information should not be available as it is relational and not intrinsic to the lexical elements involved. Put another way, the labeling one finds with adjuncts differs from that found with complements, but it is not clear how this labeling is to be interpreted. In the next sections, we will suggest that the critical difference between complements and adjuncts is that the former *requires* integration into structures with labels while the latter does not. This gives adjunction structures greater grammatical latitude, in some respects. But before discussing adjunction in detail we need to outline some principles of phrasal composition.

4.3 Labeling without bars

Let's assume a simple view of phrase structure in which adjunction is not marked by any special kind of labeling convention. Under this view an adjunction structure will look something like (14) given BPS assumptions.

(14) $[_X [_X [_X$ X YP] WP] ZP]

Given conventional assumptions, the two innermost X-marked constituents in (14) will be understood as X′s, while the outer one will be understood as an XP. In addition, it is conventionally assumed that YP can be read as the internal argument of X as it is the immediate projection of X. All these are *relational* notions and they can be defined for structures like (14) if they need to be. One place where this information may be important is at the interfaces, where syntactic configurations are interpreted. A strong version of the Inclusiveness Condition (which we are adopting here) allows such relational notions to *only* be relevant at the interfaces and not in the syntax proper, where only the intrinsic properties of lexical items are manipulated or noted.

How does the syntax "read" (14)? Chapter 3 assumes that the labels are understood conventionally (as in Chomsky 1955) via the "is-a" relation and that being bracketed together means that the bracketed elements have been concatenated. Given this, we read in (14) that X concatenated with YP (X^YP) is an X. In other words, concatenation plus labeling delivers back to one of the original concatenates. WP and ZP are read in the same way: $[_X$ X^YP]^WP is an X and $[_X [_X$ X^YP]^WP]^ZP is an X. In effect, repeated concatenation and labeling produce bigger and bigger X-objects. In each case above, YP, WP, and ZP interact with X (and only with X) via concatenation. If the CI interface understands concatenation here in terms of conjunction, then each concatenative step introduces another conjunct.[8] We will return to this point in

[8] Predicate conjunction given a Davidsonian event semantics. See Pietroski (2005) for discussion.

a moment. For now, let's consider how (14) fares with respect to the empirical properties noted in 4.2.

The fact that adjunction has no effect on selection follows directly as the head of the adjunction structure in (14) is the same as the head of a structure free of adjunctions. What is less clear is how the ellipsis, anaphora and movement operations that seem to target specific projection levels (e.g. VP ellipsis, VP fronting, *one* substitution targeting NPs, etc.) are to be reformulated given a phrase structure like (14). Let's rehearse the basic facts and see precisely what role bar-level information played before we consider an alternative.

Let's examine VP movement, for concreteness:

(15) a. It was kick Fred that John did
b. It was kick Fred that John did in the yard
c. It was kick Fred in the yard that John did
d. It was kick Fred in the yard that John did at noon
e. It was kick Fred in the yard at noon that John did
f. *It was kick that John did Fred

The paradigm in (15) can be described using bar-level information as follows: Vmaxs (but no V^n, n not max) can be clefted.[9] Modifiers adjoin to VP and the output of adjunction is bar-level identical to the input. Thus if the structure of the affected VPs in (15) is as in (16), then structure preservation constraints (conditions that require Xmaxs in specifier and complement positions) lead us to expect the pattern in (15).

(16) $[_{VP}\ [_{VP}\ [_{VP}$ kick Fred] in the yard] at noon]

In particular, the reason that *kick Fred* plus any number of adjuncts can be fronted is that *kick Fred* in (16) is a Vmax and so is *kick Fred* plus any of the adjuncts. Moreover, the reason why (15f) is unacceptable is that *kick* is not a Vmax and so structure preservation blocks its movement to a Spec position.

The problem with (14) given the paradigm in (15) is that the structure of *kick Fred in the yard at noon* would not be (16) but (17) and if we assume that bar-level information cannot be used, then it is unclear why the data distribute as seen.

(17) $[_{V}\ [_{V}\ [_{V}\ [_{V}$ kick] Fred] in the yard] at noon]

There are, to be specific, two problems with (17), one more general than the other. The more general one is how to prevent targeting *kick* for movement, as

[9] Paul Pietroski (p.c.) observes that the unacceptability of (15f) is not the result of a semantic difficulty. He observes that there is a fine meaning expressed by (15f), viz. kick(ing) is what John did to Fred . . . We might express this more colloquially as "a kick(ing) is what John gave to Fred . . ." At any rate, the problem seems not to be with the meaning expressed but with the structure exploited.

in (15f). If *kick Fred*, *kick Fred in the yard*, and *kick Fred in the yard at noon* are all Vs and can move, why can't *kick*, which is also a V, move?

The more specific problem with (17) concerns structure preservation. Recall that in Chapter 3 we derived structure preservation based on two assumptions: that morphology can only operate on lexically simple expressions and that movement must obey the A-over-A condition (A/A).[10] The former assumption is of no moment here, so we put it aside (but see 4.4 for discussion). However, the second is very relevant in at least two respects. First, we can use the A/A reasoning to explain why it is that (15f) is unacceptable. Note that the V *kick* moves out of the larger V *kick Fred*. This is an A/A violation and should not be permitted. Second, given this same reasoning, the V movements in (15b) and (15d) both violate the A/A condition and so should both be barred.

Clearly these pair of points are related and it would be nice to figure out a way to preserve the positive effects of this and hence derive the unacceptability of (15f) while at the same time figuring out why (15b) and (15d) are fine. This is what we aim to do in the next section.

4.4 No labeling

How are phrases composed? Chapter 3 proposed that there are two operations: concatenation (aka Merge) and labeling. When two elements are concatenated, one of the two marks this blessed event by giving the result its name. In (18), X and Y concatenate and X names the resulting object X.

(18) $[_X$ X^Y]

Combining Chomsky (1955) and BPS, we read (18) as saying that X concatenated with Y is (an) X. Labeling is required to derive complex embedded objects, for concatenation is defined over a set of atoms and labeling turns a non-atomic complex concatenate into a (complex) atomic element suitable for further concatenation. In other words, what labels *do* is allow concatenation to apply to previously concatenated objects by bringing these complexes into its domain (see Chapter 3 for further details). Assume that this is the correct way of construing Merge.

We can now ask whether labeling is always required after concatenation. What happens if we fail to label? In other words, how should we read (19)?

(19) $[_X$ X^Y]^Z

Here the concatenate X^Y is (an) X but not so $[_X$ X^Y]^Z. The two objects contrast in that the former is a concatenate *and an atomic object that can be*

[10] Recall that Chapter 2 shows that the A/A condition is itself an instance of minimality understood as minimizing path length.

input to further concatenations, whereas the latter is a concatenate but it *is not an atomic object* and so *cannot be input to further concatenation*. Z, as it were, dangles off the complex [$_X$ X^Y] without being integrated into a larger X-like expression. Assume that "adjuncts" can so dangle, whereas arguments must be integrated into larger structures via labeling.[11] In other words, whereas Z can be interpreted as an adjunct in (19), it cannot be interpreted as an argument. Under this view, a syntactic object such as *eat the cake in the yard* may have the structure in (20a) below, where *in the yard* is just concatenated with a projection of V, or the structure in (20b), where the result of the concatenation is also labeled as ("is a") V.[12] Furthermore, on the assumption that only labeled elements (syntactic constituents) can be targets of syntactic operations,[13] it should be possible to move *eat the cake in the yard* in (20b), but not in (20a).

(20) a. [$_V$ eat^the-cake]^in-the-yard
b. [$_V$ [$_V$ eat^the-cake]^in-the-yard]

What does this buy us? Recall that syntactic operations like VP movement can target a V+complement plus any number of adjuncts, but not a V alone, as illustrated in (21) (see (15) above).

(21) a. eat the cake he did in the yard
b. eat the cake in the yard he did
c. *eat he did the cake in the yard

If adjuncts need not resort to labeling to be licensed, as proposed here, the two possibilities in (21a) and (21b) are due to the two different structures that may underlie *eat the cake in the yard*. That is, (21a) is to be associated with (20a) and (21b) with (20b). Notice (21a) cannot be associated with (20b), for movement of *eat the cake* would violate the A/A condition as it is part of a larger V-projection. In turn, (21b) cannot be associated with (20a), for *eat the cake in the yard* is not a syntactic constituent in (20a) and therefore cannot undergo movement. More interestingly, although the structural ambiguity of *eat the cake in the yard* allows licit derivations for (21a) and (21b), it is impossible to move *eat* alone in either (20a) or (20b) without violating the A/A condition, as *eat* is a V contained within a larger V that can be target of the same operation. Thus, *if* complements must be inside labeled concatenates and adjuncts need not be, we can ascribe the unacceptability of examples like (21c) to a violation of the A/A condition.

11 This proposal is not original. It has a clear precursor in Chametzky (2000), which proposes that adjunction yields a non-labeled constituent. The proposal here is clearly a version of Chametzky's. This same idea is also developed in Uriagereka (2002).

12 We abstract away from the internal structure of the complement DP and the adjunct PP. We treat them here as atoms.

13 Because only they can be concatenated and hence be reintegrated into the phrase marker. See Chapter 3 where we outline how to restrict movement, binding, and ellipsis to constituents.

We have outlined the one adjunct case. The multiple adjunct case will function similarly. An expression such as *eat the cake in the yard with a fork in the afternoon*, for example, may have the structure in (22) below, where each PP is concatenated with the same labeled concatenate, forming a kind of "pile." Under (22), only *eat the cake* will be able to move, yielding (23), as it is the largest V-projection.

(22) $[_V$ eat^the-cake]^in-the-yard^with-a-fork^in-the-afternoon
(23) eat the cake he did in the yard with a fork in the afternoon

Alternatively, we may also have structures in which one, more than one, or all the adjuncts are integrated into a larger V-projection through concatenation and labeling, as in (24) below, for instance. Under the structures in (24), the A/A condition will enforce movement of the largest V-projection, stranding adjuncts that were added to the structure without resort to labeling, as respectively shown in (25a)–(25c).

(24) a. $[_V$ $[_V$ eat^the-cake]^in-the-yard]^with-a-fork ^in-the-afternoon
b. $[_V$ $[_V$ $[_V$ eat^the-cake]^in-the-yard]^with-a-fork]^in-the-afternoon
c. $[_V$ $[_V$ $[_V$ $[_V$ eat^the-cake]^in-the-yard]^with-a-fork]^in-the- afternoon]

(25) a. eat the cake in the yard he did with a fork in the afternoon
b. eat the cake in the yard with a fork he did in the afternoon
c. eat the cake in the yard with a fork in the afternoon he did

Again, neither (22) nor structures like (24) allow movement of the verb alone without violating the A/A condition; hence the unacceptability of (26).

(26) *eat he did the cake in the yard

To sum up the discussion thus far: A labeled concatenate is a complex atom. Atoms have no accessible innards. By rendering a complex concatenate atomic, the label prevents the insides of the labeled elements from being targets of movement by the A/A condition.[14] When adjuncts don't move with the elements they modify, it is because they are not members of the *labeled* concatenate that has moved (cf. (24)/(25)). However, arguments can never be other than members of a labeled concatenate, for their interpretive lives depend on it. A side effect of this requirement is that heads that theta-mark complements become ineligible

14 This reasoning extends to *one* substitution cases and ellipsis on the assumption that A/A is respected here, as well. Chapter 3 shows that if ellipsis involves movement (as suggested in Johnson 2001) then the A/A should naturally apply. Similarly if *one* substitution involves movement, as occurs with the analogous *ne* construction in Italian. The logic above is further compatible with proposals that consider *one* to be thematically inert (unable to assign a theta-role, see Panagiotidis 2003). If so, having *one* as an anaphoric head prevents its complement from integrating into the proposition ((10) above). The same account presumably can extend to the *do so* cases if this is seen as the verbal counterpart of *one* ((9) above).

targets as the derivation of sentences such as (26) is ruled out by the A/A condition.

The astute reader (and what other kinds are there?) will have noted that this is not entirely satisfactory. We need an explanation for *why* there is this distinction between arguments and adjuncts, for otherwise haven't we simply recoded the facts? Though I agree that an explanation is needed (and I will try to provide one in a moment), it behooves us to note that if the above is tenable, then we have already accomplished something. We have attributed the label properties of adjunction constructions to structural ambiguity rather than to a novel labeling convention. What distinguishes adjunction structures is not a new *kind* of label but the absence of one. The V+complement in the non-labeled adjunction structure is clearly maximal for nothing with a different label dominates it in the relevant configuration. Where the V+complement plus a number of adjuncts move, the V+complement is not maximal. When the V+complement+adjuncts moves, it is this V+complement+adjunct that is the maximal V. In other words, there is nothing amiss about labeling the whole moving constituent a projection of V in just the way that V+complement is a labeled projection of V. In other words, once one allows adjuncts to live within non-labeled concatenates, the standard facts about adjuncts are accommodated without running afoul of BPS conceptions. Clearly, more needs to be said about structures such as (22) or (24).[15] However, this is sufficient detail for the time being.[16]

Let's now turn our analytical gaze to head adjunction structures.[17] Take V-to-T movement, for concreteness. If we were to translate the standard Chomsky-adjunction structure in (27) below in terms of the proposal advocated here, we should get something along the lines of (28), with T concatenating with V twice. In one case, this yields a labeled constituent and in the other case, it doesn't.

[15] For instance, one must determine the interface conditions that presumably motivate/license labeling in structures such as (24). See below for some discussion. Also, linearizing adjunction structures such as (22) and (24) appears to require special provisos. Chomsky (1995a) argues that adjunction is unlikely to fall under the purview of the LCA. If so, then the linearizing adjuncts will require special considerations on any theory of adjunction.

[16] This proposal might be taken as fleshing out the oft-mentioned idea that adjuncts inhabit another dimension (see Chomsky 2001) while making it compatible with a single cycle theory. Given the absence of LF in single cycle theories, the integration of adjuncts cannot wait until LF to "integrate" them. Concatenating adjuncts without labeling them might be construed as having them in "another dimension" without leaving them entirely outside the phrase marker. At any rate, it appears that LF operations like pronominal binding and obligatory control are only licit within single rooted phrase markers. Hence if the required integration does not take place at LF, it must take place in overt syntax and the distinction between concatenated and labeled versus merely concatenated serves to make the necessary distinction. For further discussion of single cycle theories see Chapter 6.

[17] Assuming that head movement exists in the grammar. See Chapter 3 for discussion of the various alternatives. It would greatly simplify matters if head movement failed to obtain in UG. What follows is relevant on the assumption that head movement is a licit operation.

(27)

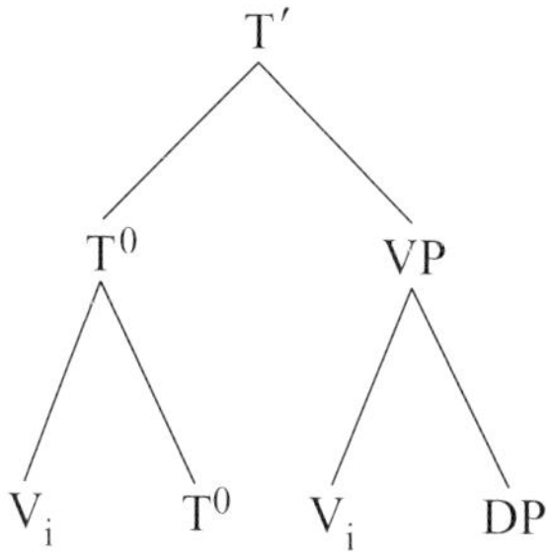

(28) $[_{T}\ T\hat{}[_{V}\ V\ D]]$
$\hat{}V$

Structures such as (28) raise several questions. First, why isn't the first merger between T and a projection of V sufficient to establish all the necessary relations between T and V? That is, why must T merge with (a projection of) V twice? Second, movement of the V-head appears to violate the A/A condition, given that it is dominated by a larger V-projection. Third, when V concatenates with T for the second time, it does not target the root of the tree, thus violating the Extension Condition (Chomsky 1995a). Finally, head adjunction structures do not behave like XP-adjunction structures with respect to the movement possibilities. Descriptively speaking, XP-adjunction structures allow the adjunct and the target of the adjunction to move independently of one another. By contrast, in head adjunction structures movement of the adjoined element ("excorporation") is taken to be impossible (Baker 1988) or severely restricted (Roberts 1994). Moreover, it seems to be a point of consensus that the head of an adjunction structure cannot be excorporated, leaving the adjoined head stranded.

Let's consider two approaches under which head-to-head movement would be compatible with our proposal. Under the first approach, the problems reviewed above are not real because head movement is actually a PF phenomenon and not part of narrow syntax (see Boeckx and Stjepanović 2001 and Chomsky 2001: 38, among others). If this approach is correct, the problems above actually provide a rationale for this gap in the computations of narrow syntax.

Under the second approach, the problems are real, but tractable. A common assumption within minimalism is that if an expression X assigns a theta-role to Y, then it cannot also check a feature, say Case, of Y (see Chomsky 1995a, Grohmann 2003). So, for example, a "transitive" light verb assigns a theta-role to its Spec, but checks the Case-feature of the DP that is theta-marked by the lower verb. In other words, the assumption is that the one and the same head cannot simultaneously theta-mark and morphologically check the same expression. One could extend this division of labor to other morphological relations, as well. So, if T has both morphological and selection requirements to be satisfied by V, T must concatenate with (a projection of) V twice. Furthermore, it is arguable that morphological requirements must involve simplex (word-like) elements and not complex atomic elements (phrases).

If this is the case, the A/A condition should be understood in a relativized manner, viz. if a complex element such as the labeled projection $[_V$ V D$]$ cannot satisfy the morphological requirements of T (it is not word-like), it does not induce minimality effects of the A/A type for the movement of the simplex verbal head (see Chapter 3 for discussion). From this perspective, excorporation of the adjoined head (e.g. V) or the target of adjunction (e.g. T) will plausibly cause the derivation to crash for several reasons. First, it is plausible that excorporating V will result in a morphologically ill formed "word," as the affix will no longer be supported. Further, if T projects and labels the V^T concatenate it will block movement of the T head by the A/A condition. This labeling also forces the adjoined V to pied-pipe with the moving T.[18] And like the previous V-to-T movement, if $[_T$ V^T$]$ moves for morphological reasons, the larger complex projections of T will be inert for purposes of the A/A condition. Finally, if we assume that head movement is subject to the Extension Condition then head movement must be an instance of sideward movement (see Bobaljik 1995b, Bobaljik and Brown 1997, Nunes 1995, 2004, and Uriagereka 1998). That is, the verb must be copied from within $[_V$ V^D$]$ and concatenated with T *prior* to the merger between T and $[_V$ V^D$]$, as illustrated in (29). Lastly, we can force labeling under head movement if we assume that morphological processes only apply to lexical items. An unlabeled V^T is not a lexical item, derived or otherwise. $[_T$ V^T$]$ is. If this holds then head movement requires labeling on pain of morphological uninterpretability. With this assumption the derivation in (29c) would be ill formed and replaced by the one in (29d,e).

(29) a. *Assembly of* $[_V$ *V^D*$]$ *+ selection of T from the numeration*: $[_V$ V^D$]$ T
b. *Copy of V from* $[_V$ *V^D*$]$ *+ Concatenation with T:* $[_V$ V^D$]$ T^V
c. *Concatenation of T with* $[_V$ *V^D*$]$ *+ labeling (cf. (28)):*
$[_T$ T^$[_V$ V D$]]$ ^V
d. *Copy of V from* $[_V$ *V^D*$]$ *+ Concatenation with T + labeling by T:*
$[_V$ V^D$]$ $[_T$ T^V$]$
e. *Concatenate sub-trees in (d) and label with T:* $[_T$ $[_T$ T^V$]$^$[_V$ V^D$]]$

OK, we have dallied long enough: why the labeling differences between adjuncts and complements? What *conceptually* motivates the different treatment that we have seen is empirically required? We believe that the proposed difference tracks an independently required semantic contrast between the

[18] In this case, the resulting structure would be as in (i).

(i) $[_T$ $[_T$ V^T$]$^$[_V$ V^D$]]$

two, namely the fact that to be predicated of events, arguments (in contrast to adjuncts) need a thematic pivot. Here's what we mean.

In a neo-Davidsonian semantics the core of the proposition is the event.[19] The V is a predicate of events and everything else modifies it. Thus, the logical form of (30a) is something like (30b).

(30) a. John ate the cake in the yard
b. ∃e [eating(e) & subject(John,e) & object(the cake, e) & in-the-yard(e)]

The crucial feature of (30b) for current purposes is that the verb *eat* and the adjunct *in the yard* apply to the event directly, whereas *John* and *the cake* modify the event via two designated relations, here marked "subject" and "object." Whether it is grammatical functions like external/subject and internal/object or thematic relations like agent and theme/patient is irrelevant here. What is important is that adjuncts can *directly* modify events, while arguments only do so *indirectly*. They need help in relating to the event and this help is provided by relational notions like subject, object, etc. In an event-based semantics, arguments – not adjuncts – are the interpretive oddballs. They can only modify the event if aided by relational notions.

How does this bear on the requirement that arguments must be inside labeled concatenates while adjuncts need not be? If we assume the traditional definitions of "subject," "object," etc., then we need labels.[20] For example, objects are traditionally defined as the immediate concatenates of V, e.g. NP-of-V/[$_{VP}$ V NP] in the Standard Theory. Given the assumptions that the object/subject relation must be marked so as to be of use at the CI interface (the place where the syntactic object is interpreted, viz. integrated into a neo-Davidsonian event-based proposition), we must provide the structural wherewithal to define it. And, if we understand notions like subject and object in classical terms, then labeling is critical for defining these relations. Thus, whereas arguments necessarily require being in a complex labeled structure, adjuncts can be licensed with simple concatenation.

Assuming that this proposal is on the right track, let's consider some of its implications for the computation of adjuncts.

4.5 Some consequences

The traditional description of adjunction structures is that the adjunct somehow dangles off the target of adjunction. This accounts for the fact that when the target moves as in VP-fronting, for instance, it may pied-pipe the adjunct or

19 For details, see Higginbotham (1986), Parsons (1990), Pietroski (2005), and Schein (1993), among others.

20 See, for example, Chomsky (1965).

leave it stranded (cf. (23) and (25)). We have reanalyzed this optionality in terms of structural ambiguity. When the adjunct is left stranded, that's because its concatenation with the target was not followed by labeling, as sketched in (31a); on the other hand, if the adjunct is carried along, labeling has taken place, as represented in (31b).

(31) a. $[_V$ V^D]^Adj
b. $[_V$ $[_V$ V^D]^Adj]

In this section we will focus on structures such as (31a). Assuming that concatenation without labeling is a grammatical possibility for adjuncts, the structure in (31a) invites two inferences. On the one hand, the adjunct should be invisible to operations involving the labeled structure, as it is "dangling off" the labeled V. On the other hand, given that it is not dominated by a labeled structure, the Extension Requirement does not prevent it from merging with another element. That is, the adjunct in (31a) may "dangle onto" a different structure. We discuss each possibility below.

4.5.1 Dangling off

One finds evidence from different domains that indicates that adjuncts may be invisible to certain grammatical computations. For instance, as opposed to arguments, adjuncts do not project focus (see Gussenhoven 1984, Selkirk 1984, among others). A sentence such as (32a), for example, with *car* being prosodically prominent, can be a felicitous answer to *What did John buy?* (object focus), *What did John do?* (VP focus), or *What happened?* (sentence focus). By contrast, a similar sentence with a prosodically salient adjunct such as (32b) can only be an appropriate answer for *Where does John read books?* (adjunct focus).

(32) a. John bought a CAR
b. John reads books in the CAR

From the perspective explored here, the contrast between arguments and adjuncts with respect to focus projection is a by-product of the fact that arguments must be fully integrated into their structure (concatenation and labeling are both required), whereas adjuncts are allowed to be dangling out (only concatenation is required), as shown in the simplified structures in (33). In other words, as arguments necessarily become integral parts of larger and larger labeled structures, they allow focus to project to these structures; in turn, as adjuncts are just concatenated, they are not very communicative with their neighbors. In effect, this is to assume that only a labeled node can project focus. The non-labeled node that results from just concatenating the adjunct (and not labeling the result) is insufficient to project focus further and thus restricts it to the adjunct.

(33) a. $[_T$ John^$[_T$ T^$[_V$ bought^a-CAR]]]
b. $[_T$ John^$[_T$ T^$[_V$ reads^books]]] ^in-the-CAR

The contrast in (32) supports two observations. First, it shows that labeling is not optional. If it were, the concatenate in (33b) could be labeled and the distinction between arguments and adjuncts with respect to focus projection would be lost. Second, if labeling concatenated structures involving adjuncts is not optional but must be triggered by some interface conditions (see note 15), focus projection is not one of them. If it were, it would license the labeling in (33b) and, again, we would have no principled basis to account for the different behavior of arguments and adjuncts regarding focus.

Say this is on the right track. Doesn't it contradict our proposal in 4.4 that the multiple choices for VP movement rested on structural ambiguity, depending on whether or not a concatenate involving an adjunct is labeled? Not really. To say that a given surface string involving multiple adjuncts may correspond to different structural configurations that depend on whether the concatenation of the adjuncts was followed by labeling does *not* entail that labeling is optional. All that it entails is that whatever triggers/licenses labeling in these cases must have been enforced when adjuncts are pied-piped under VP movement.[21] Our proposal in fact predicts that, all things being equal, adjuncts should be able to project focus once the labeling is properly sanctioned. In other words, an adjunct should be able to project focus if pied-piped in a fronted VP.

With this in mind, consider the contrast in (34).

(34) [Context: *What will John do?*]
a. #He will play soccer on SUNDAY
b. Play soccer on SUNDAY is what he'll do

As mentioned above, a question such as *What will John do?* can be used as a diagnostics for VP focus and, therefore, the sentence in (34a) with high pitch on *Sunday* is expected to be infelicitous, as it only licenses narrow focus, i.e. it would only be a felicitous answer to the question *When will John play soccer?* Interestingly, the corresponding pitch accent on "SUNDAY" with VP fronting under pseudoclefting in (34b) is a suitable answer in the context given. From the perspective of our proposal, the fact that the adjunct is pied-piped in (34b) signals that labeling after concatenation was licensed. Once fully integrated into the structure, focus can then propagate from the adjunct to the larger VP

[21] If VP movement underlies VP ellipsis as suggested by Johnson (2001) and Szczegielniak (2005) and reviewed in Chapter 3 then several possibilities available for ellipsis involving multiple adjunction should fall together with VP fronting, as far as the licensing of labeling involving the concatenation of adjuncts is concerned.

of which it becomes an integral part. Thus, even though the exact trigger for such labeling remains to be specified, the contrast in (34) lends support to our account of the general asymmetry between arguments and adjuncts with respect to focus projection in terms of (lack of) labeling.

It is worth observing that leaving a structure unlabeled is more economical (in the sense that fewer operations are applied) than labeling it. Thus, according to general minimalist logic, it should not occur unless required. What requires it? Regular sentence intonation suffices to focus the VP. Additional pitch accent, even on the V, shifts focus from the VP as a whole: *What did John do? # He PLAYED soccer on Sunday*. Thus, if VP focus is intended, no labeling is required and so none should occur. Moreover, focusing the PP does not require integrating the PP into a labeled structure. What we see in (34b) is the whole VP being focus-moved but the pitch accent on the complement of a PP adjunct. In other words, here pitch accent and focus do not track one another. What allows this to happen, on the view presented here, is that focus-moving the whole VP-plus-adjunct requires labeling it as V and this licenses a pitch accent consistent with the VP focus interpretation even though this same intonation would not license this same interpretation without movement. This suggests that labeling in (31b) is only grammatically available if the VP-plus-adjunct is moved (and possibly subsequently elided). If there is no movement, only the labeling in (31a) is licit.[22]

Consider another domain in which adjuncts appear oblivious to the computations in play. As illustrated by the contrast in (35), for instance, the negative head *not* blocks affix hopping (see Chomsky 1957), but the adjunct *never* doesn't.

(35) a. *John not baked cakes
b. John never baked cakes

The contrast above receives a straightforward account under the standard assumption (see Pollock 1989) that *not* heads a labeled constituent (NegP) intervening between T and VP, whereas the adjunct *never* is just concatenated with VP, as respectively shown in (36). Crucially, *never* is dangling off of VP in (36b) and does not interfere with the adjacency requirements on affix hopping (see Bobaljik 1995a for discussion).[23]

[22] There appear to be other intonational differences between fronted adjuncts and those in base position. For example, in examples with multiple adjuncts (*John ate the cake in the gazebo, with a fork, at noon, in the rain*) the adjuncts show a flat list like cadence. This contrasts with the cadence observed when the whole large VP is fronted. See Wagner (2005) for relevant discussion.

[23] See also Avelar (2004), who argues that different arrangements among the functional heads *v*, T, D, Poss, and Top in Brazilian Portuguese underlie the lexical access to the copulas *ser* "be" and *estar* "be" and the existential/possessive verb *ter* "have." Interestingly, "intervening" adjuncts are disregarded and do not interfere with the access to a particular vocabulary item.

(36) a. [$_{\text{T}}$ -*ed*^[$_{\text{Neg}}$ not^[$_{\text{V}}$ bake^cakes]]]
b. [$_{\text{T}}$ -*ed*^[$_{\text{V}}$ bake^cakes]] ^never

This proposal may also underwrite an account of some unorthodox aspects of grammatical computations when adjuncts are involved. Take the standard assumption that syntactic operations do not target discontinuous elements, for instance. When cases such as (37) and (38) below are considered, it seems that this requirement must be relaxed as far as adjuncts are concerned, for VP movement, ellipsis and *do so* anaphora appear to be targeting a discontinuous object (*eat the cake in the afternoon* in (37) and *eat the cake with a fork* in (38)).

(37) John ate the cake in the yard with a fork in the afternoon
a. . . . and eat the cake in the afternoon, he should have in the kitchen, with a spoon
b. . . . but Bill did (so) in the kitchen, with a spoon

(38) John ate the cake in the yard with a fork in the afternoon
a. . . . and eat the cake with a fork, he should have in the kitchen in the morning
b. . . . but Bill did (so) in the kitchen in the morning

However, the fact that adjuncts can be left dangling provides an alternative analysis of data such as (37) and (38), which is compatible with the standard assumption that discontinuous objects cannot be targeted by syntactic operations. Recall that 4.4 proposed that it is structural ambiguity that allows VP movement, ellipsis, and *do so* anaphora to also target any number of adjuncts without violating the A/A condition. The idea is that the multiple possibilities for these grammatical operations are actually associated with different syntactic structures, depending on whether or not concatenation of the adjuncts is followed by labeling. The same can be said about the sentences above. That is, (37) is to be associated with the structure in (39), and (38) with the one in (40).

(39) [$_{\text{V}}$ [$_{\text{V}}$ ate^the-cake]^in-the-afternoon]
^in-the-yard
^with-a-fork

(40) [$_{\text{V}}$ [$_{\text{V}}$ ate^the-cake]^with-a-fork]
^in-the-yard
^in-the-afternoon

Given the structures in (39) and (40), the object that is targeted by the computational system in (37) and (38) is indeed a labeled concatenate (a syntactic atom) and not a discontinuous element. Rather than requiring some relaxation in the computational system, what sentences such as (37) and (38) actually do is show that the surface order among the adjuncts does not provide any information as to whether or not labeling has occurred. Or, more succinctly, the linearization of adjuncts in the PF component does not seem to be ruled by the same mechanisms that deal with the linearization of arguments (see note 15).

There is an additional happy consequence. Regardless of whether ellipsis resolution is to be ultimately accounted for in terms of PF deletion or LF copying, we have seen that ellipsis in (37a) and (38a) arguably disregards adjuncts that were merely concatenated into the structure. This opens a new avenue for the analysis of ellipsis resolution that may lead to infinite regress such as the ones in (41).

(41) a. John greeted everyone that I did
b. John worded the letter as quickly as Bill did
c. John kissed someone without knowing who

(41a) is a classical example of antecedent contained deletion (ACD) construction of the sort first extensively discussed in May (1985). (41b) is an ACD construction in which the major constituent containing the elided material is an adjunct (see Hornstein 1995). Finally, (41c) involves sluicing contained within an adjunct (see Yoshida 2006). In all of them, a simple-minded ellipsis resolution copying the matrix VP in (41a) and (41b) or the IP in (41c) into the ellipsis site will recreate a structure with elided material in need of resolution. This is not the place to discuss the intricate properties associated with each of these constructions. I would just like to point out that they appear to be amenable to the same analysis I suggested for (37a) and (38a).

More concretely, the infinite regress problem arises just in case the adjuncts in (41) are analyzed as forming a syntactic constituent with the target of the adjunction. Suppose that along the lines we have been exploring here, the simplified structures underlying the sentences in (41) are as in (42).

(42) a. $[_{T}$ John^$[_{T}$ T^$[_{V}$ greeted^everyone]]]
^that-I-did
b. $[_{T}$ John^$[_{T}$ T^$[_{V}$ worded^the-letter]]]
^as-quickly^as-Bill-did
c. $[_{T}$ John^$[_{T}$ T^$[_{V}$ kissed^someone]]]
^without-knowing-who

In each structure of (42) there is a constituent that can provide the relevant template for ellipsis resolution without forcing infinite regress; namely, the V-labeled concatenate in (42a) and (42b) and the outer T-labeled concatenate in (42c). The crucial aspect in the structures in (42) is that the adjunct containing the ellipsis site is just concatenated with its target and therefore is not a proper part of the structure it modifies. As it dangles off the constituent with which it was concatenated, it is invisible for purposes of ellipsis resolution and this doesn't lead to the infinite regress trap.[24]

The purpose here has been to highlight empirical domains that may find a more streamlined explanation if our proposal that adjuncts may be just concatenated with their target is on the right track. Clearly, these cursory remarks do not provide a sufficiently detailed analysis of the several types of phenomena reviewed in this section.[25]

4.5.2 *Dangling on*

There is one more aspect of adjunction structures that we haven't mentioned here. Grammarians distinguish between domination and containment (see May 1985). According to this distinction, XP in (43a) below is in the domain of Y^0 but not in the domain of Z^0 as it is dominated by all maxY projections. In contrast, XP in (43b) is in the domain of both Y^0 and Z^0 because it is not dominated by all maxY projections; that is, it is dominated by ZP but only contained by YP.

(43) a. $[_{ZP} \ldots Z^0 [_{YP}$ XP $[_{Y'} \ldots Y^0 \ldots]]]$
b. $[_{ZP} \ldots Z^0 [_{YP}$ XP $[_{YP} \ldots Y^0 \ldots]]]$

The distinction between domination and containment has been empirically useful in allowing expressions to be members of more than one domain. One case that illustrates this possibility is provided in Kato and Nunes's (1998) analysis of matching effects in free relatives. In Portuguese, for example, free relatives allow a kind of preposition sharing between different verbs. The data in (44) below show that the verbs *discordar* "disagree" and *rir* "laugh" in Portuguese select for the preposition *de* "of," whereas the verbs *concordar* "agree" and *conversar* "talk" select for the preposition *com* "with." When one of these verbs takes a free relative clause as a complement, the selectional properties of the matrix and the embedded verb must match, as shown in (45). Intuitively speaking, (45c), for instance, is ruled out because the matrix verb selects for *com*, while the embedded verb selects for *de*:

(44) a. Eu discordei/ri **d**ele /***com** ele
I disagreed/laughed of-him with him
"I disagreed with him."/"I laughed at him."

[24] See Nakao (2007) for an analogous proposal.

[25] If movement is to be computed in terms of paths (see Chapter 2) and if paths are defined in terms of traversed constituents (labeled concatenates in our terms), lack of labeling might block movement if paths can't be computed. In other words, lack of labeling might provide a partial account for why one can't move out of adjuncts. If something along these lines is correct, it remains to be explained why moved adjuncts are also islands. See Chapter 7 for some further discussion.

b. Eu concordei/conversei **com** ele /***d**ele
I agreed talked with him of-him
"I agreed with him."/"I talked to him."

(45) a. Ele só conversa **com** quem ele concorda.
he only talks with who he agrees
"He always talks to who he agrees with."

b. Ele sempre ri **de** quem ele discorda
he always laughs of who he disagrees
"He always laughs at who he disagrees with."

c. Ele sempre concorda ***com** quem/***de** quem ele ri
he always agrees with who of who he laughs
"He always agrees with who he laughs at."

d. Ele sempre ri ***de** quem/***com** quem ele conversa
he always laughs of who with who he talks
"He always laughs at who he talks to."

Assuming the traditional distinction between domination and containment, Kato and Nunes propose that the derivation of a sentence such as (45a), for instance, proceeds as follows. The computational system assembles the "relative" CP and the verb *conversa* is selected from the numeration, as shown in (46) below. K and L in (46) cannot merge at this point because *conversa* does not select for a CP. The strong *wh*-feature of C then triggers the copying of the PP *com quem*, as shown in (47). Next, the computational system adjoins M to K, allowing the strong *wh*-feature to be checked, and merges the resulting structure with L, as shown in (48). Crucially, merger of the matrix verb and CP in (48) now satisfies Last Resort because the moved PP also falls within domain of *conversa* and they can establish the relevant syntactic relation (theta-assignment).

(46) a. K = [$_{CP}$ C [ele concorda [$_{PP}$ com quem]]] *(he agrees with who)*
b. L = conversa *(talks)*

(47) a. K = [$_{CP}$ C [ele concorda [$_{PP}$ com quem]$_i$]] *(he agrees with who)*
b. L = conversa *(talks)*
c. M = [$_{PP}$ com quem]$_i$ *(with who)*

(48) [$_{VP}$ conversa **[$_{CP}$** [$_{PP}$ com quem]$_i$ **[$_{CP}$** C [ele concorda [$_{PP}$ com quem]$_i$]]]]
talks with who he agrees with who

In sum, the utility of distinguishing containment from domination is that elements contained within a projection are still visible beyond that projection, while those dominated by a projection are not. However, this distinction crucially hangs on allowing XP in a structure like (43a) to be distinguished from XP in a structure like (43b) and this brings back all the questions we discussed in 4.2. Note, for instance, that the assumption that the lower YP in (43b) determines the label of the outer projection but retains its status as a maximal projection is at odds with the notion of projection in BPS. In addition, it violates the Inclusiveness Condition in that bar-level information is tacitly being used as a primitive by the computational system. Moreover, notice that if these problems were to be fixed in consonance with BPS and the Inclusiveness Condition, (43b) should be reanalyzed along the lines of (49) below, where bar-levels are not intrinsically distinguished. The problem now is that we lose the distinction between adjuncts and specifiers that was used to account for the matching effects in (45), for (49) would be the BPS rendition of both (43a) and (43b).

(49) $[_Z \ldots Z\ [_Y X\ [_Y \ldots Y \ldots]]]$

The question before us is whether the apparently useful distinction between domination and containment can be captured in a theory that does not have specific labels for adjuncts such as the one we are advocating here. If so, it could evade the above noted difficulties with BPS and the Inclusiveness Condition. Consider the following: Recall that above we suggested that adjuncts can Concatenate with concatenative atoms and that the result need not project a label. Given this, we can represent the difference between domination and containment as the difference between (50a) and (50b).

(50) a. $[_X Z\hat{}[_X \ldots X \ldots]]$
b. $Z\hat{}[_X \ldots X \ldots]$

In (50a), Z has concatenated with the "inner" X-projection and the result has been labeled X again. (50b) exhibits a similar concatenation but the result is left *unlabeled*. If we assume that it is labeling that prevents all but a head to be "seen" from outside the concatenate, then in (50b) Z can still be input to further concatenation.[26]

To put it somewhat differently, recall that in 4.5.1 we discussed cases where adjuncts are disregarded by some operations because, like Z in (50b), they are not part of a labeled constituent. Once an adjunct may be left dangling as in (50b), the converse situation may arise, as well. That is, the adjunct in (50b) may be targeted by some operation exactly because it is not subpart of a bigger syntactic object. In particular, it is free to undergo merger (consistent with Extension) as it is still a syntactic atom for purposes of concatenation.

Consider how our reworked version of the distinction between domination and containment operates in the case of the Portuguese free relatives described above. The derivation of the matching free relative in (45a), for instance, can be derived along the lines of (51).

[26] Chapter 3 proposes that the Endocentricity Condition follows from a proper understanding of labeling.

(51) a. com-quem^[$_C$ C^[$_T$ . . .]]
with who

b. com-quem^[$_C$ C^[$_T$ . . .]]
[$_V$ conversa^]
talks with who

In (51a) *com quem*, which was copied from within CP, concatenates with CP and no labeling takes place. Once *com quem* is still an atomic element for purposes of concatenation, it can merge with the verb *conversa*. However, in order for *com quem* to be interpreted as an argument, such concatenation must be followed by labeling, as shown in (51b). *Com quem* in (51b) counts as two beads on a string, so to speak: it is an integral part of the V-labeled expression and a "mere" concatenate to the C-labeled expression. If one assumes that Merge is just an instance of concatenate, then there is no reason why *some* parts of the phrase marker may not be "string-like." Our suggestion is that this more adequately describes what happens for contained expressions. They are parts of "mere" concatenates, not labeled ones.[27]

[27] At first sight, our analysis fails to account for the acceptability of Portuguese sentences such as (i), for instance, where the free relative appears to have moved from the matrix object position. According to the derivation discussed above, such movement should not be possible, given that the PP and the "relative" CP do not form a constituent (cf. (51b)).

(i) Com quem ele conversa ele concorda
with who he talks he agrees
"Whoever he talks to, he agrees with."

However, upon close inspection there is a convergent source for (i), along the lines of (ii)–(vii) below (with English words and details omitted for purposes of exposition). That is, after K and L are assembled in (ii), the computational system copies *with who* and merges it with *talks* (an instance of sideward movement) to satisfy the theta-requirements of the latter (see Nunes 2001, 2004), yielding (iii). After the stage in (iv) is reached, another copy of *with who* is created, triggered by the strong feature of the Top head, as shown in (v). But before this happens, the "relative" CP may then adjoin to the copy just created (i.e. no labeling obtains after they concatenate), as shown in (vi). Given that *with who* is still an accessible atom for purposes of structure building, it may then merge with the Top-labeled constituent, yielding another Top projection, as shown in (vii), which surface as (i) and further computations. See Nunes (2001, 2004) for discussion of similar derivations.

(ii) K = [$_{Top}$ Top^[$_T$ he-agrees-[$_P$ with^who]]]
L = talks

(iii) K = [$_{Top}$ Top^[$_T$ he-agrees-[$_P$ with^who]$_i$]]
M = [$_V$ talks^[$_P$ with^who]$_i$]

(iv) K = [$_{Top}$ Top^[$_T$ he-agrees-[$_P$ with^who]$_i$]]
N = [$_C$ he-talks-[$_P$ with^who]$_i$]

(v) K = [$_{Top}$ Top^[$_T$ he-agrees-[$_P$ with^who]$_i$]]
N = [$_C$ he-talks-[$_P$ with^who]$_i$]
O = [$_P$ with^who]$_i$

(vi) K = [$_{Top}$ Top^[$_T$ he-agrees-[$_P$ with^who]$_i$]]
P = [$_P$ with^who]$_i$^[$_C$ he-talks-[$_P$ with^who]$_i$]

(vii) Q = ^[$_C$ he-talks-[$_P$ with^who]$_i$]
[$_{Top}$ [$_P$ with^who]$_i$^[$_{Top}$ Top^[$_T$ he-agrees-[$_P$ with^who]$_i$]]]

4.6 Concluding remarks

Adjuncts are funny characters from a syntactic point of view, because they appear to be simultaneously inside and outside a given syntactic tree. Their double personality has led to the standard view in the literature according to which structures involving adjuncts are less trivial than the ones involving arguments. This chapter has proposed that contrary to the traditional wisdom, exactly the opposite is true. Arguments – in order to be interpreted as such at the CI interface – necessarily require being associated to relational notions such as "subject" and "object" and the establishment of these relational notions is achieved through labeling. Hence, arguments must be part of complex (labeled) structures. Adjuncts, on the other hand, may modify the event directly via concatenation and therefore need not invoke labeled structures to be properly interpreted. From this perspective, the addition of adjuncts into a given structure is achieved via the simplest possible operation. The distinction between arguments and adjuncts, then, is conceptually based on their distinctive roles at the CI interface. Moreover, it accords well with both BPS (as it eschews use of bar-level information) and the Inclusiveness Condition (as it doesn't introduce extraneous devices to code their differences). Rather, the proposal builds on treating Merge as a species of concatenation and the idea that labeling is an operation that allows complex concatenates to further concatenate. So analyzing Merge leaves room for treating adjunction as simple concatenation without labeling (as first proposed by Chametzky) and is further motivated by a neo-Davidsonian perspective on the semantics of complements and adjuncts.

5 The emerging picture: Basic operations, FL and the Minimalist Program

5.1 Introduction

Hauser, Chomsky and Fitch (2002) have put the following question on the research agenda: what features of the faculty of language (FL) are unique and which are common across cognitive domains or reflect general principles of computation? The answer to this question is of interest to linguists for the light that it can shed on Darwin's Problem, i.e. the logical problem of language evolution. As outlined in Chapter 1, there is a tension between the distinctiveness of the basic features of FL and the apparently short time course of its emergence. The rapidity of FL's emergence suggests that it is only modestly different from non-linguistic aspects of cognition. If this is correct, then one aim of theoretical syntax should be to show how the gross features of FL result from the combination of general cognitive operations and principles plus a very small number of innovations (preferably one) specific to FL. This chapter aims to examine the proposals outlined in the preceding three chapters in this light. The chapter is short as it relies on the conclusions of Chapters 2 through 4. However, the point of the exercise is to sufficiently annotate the logical problem of language evolution so that its high level empirical implications gain a modest heft.

5.2 The basic operations of FL and the "laws" of UG

The previous three chapters have relied on the following inventory of basic operations: Concatenate, Copy, and Label. In addition, we have adopted a minimality principle that requires that dependency length of grammatical relations be measured by the size of the paths between them with the aim of minimizing this length. We have also adverted to the Inclusiveness Principle to limit grammatical operations to those involving intrinsic features of lexical items. The latter plausibly implies something akin to a Bare Phrase Structure (BPS) approach to phrase labels, a perspective we have embraced. What sort of grammars does this combination of basic operations and principles yield? What, in other words, would we expect FL to look like (e.g. what kinds of laws would we expect it to follow) if it were restricted to this basic inventory?

First, this basic inventory limits how grammatical relations are established. The only way to establish grammatical relations is by concatenation as Concatenate is the only operation that pairs expressions. The restriction to Merge as the vehicle for establishing grammatical relations is regularly assumed to be the case for relations like selection, subcategorization and theta-marking. For these relations, $\alpha R \beta$ only if α merges with β. If Merge is a species of concatenation then this carries over to the present proposals. What is less standard is the assumption that this requirement extends to *all* grammatical relations. Thus, case assignment, binding, and control similarly require concatenation between the interacting expressions. In order for this to obtain then some sort of additional operation is required. Since the earliest days of the Minimalist Program, Move has been that operation. This proposal endorses that one. Copy serves to create tokens of an expression that concatenate with the relevant case assigner, antecedent, or controller to license the relevant relation.[1] Not surprisingly, in this sort of system, displacement is expected to be a common feature of natural language grammars (as indeed it is). It will arise whenever an expression must relate to two different heads, for example (e.g. Wh-movement where the Wh-element needs a theta role, a case and a Wh-feature checked/assigned).[2] Moreover, given the assumption that Merge obeys the Extension Condition, we expect all relations to be licensed under c-command, at least in the standard cases of relations established within single-rooted subtrees. Thus, antecedents should c-command their anaphoric dependents, controllers should c-command their controllees, and displaced elements should c-command their points of origin. This, observe, is because c-command is a necessary feature of any account in which Copy, Concatenate and Extension are basic operating principles of

[1] Extending this to bound pronouns might involve de-concatenation. See Hornstein (2006) for discussion in the context of pronominalization. The present story is also consistent with treating pronoun binding via Move as in Kayne (2002) as well.

[2] Let me repeat something noted in Chapter 3 and discussed again in Chapter 7. Movement is required if an expression must fulfill grammatical requirements that cannot be checked by a single head. So, for example, in MP α cannot assign both a theta role and a case to β. Thus, if β needs both a theta role and a case then it must move so as to concatenate with both α and some case assigner γ. In contrast in GB V can both assign a theta role to its object and check its case and so movement is not required. This generalizes: if some head could, for example, check the theta, case and Wh features of a DP then movement for case and Wh feature checking would be unnecessary. This possibility is apparently not realized in natural language. Rather, heads are, by and large, restricted to checking/assigning one relevant feature per relatum. Given this, movement is inevitable. The real question then is why this is how things are organized. See Chapter 7 for some discussion.

FL. In this sense, c-command is an emergent property of the system being investigated here.[3]

Second, grammatical relations will be subject to two kinds of locality restrictions. The first is a minimality requirement in the sense of Rizzi (1990). The version of Minimality proposed here minimizes dependency length as measured by paths using Boolean yardsticks. A consequence of this is the c-command restriction on relevant interveners. Another is the A-over-A condition, which reduces to an instance of Minimality. A third is the exemption from Minimality restrictions for elements moving within or to the same minimal domain (in the sense of Chomsky 1995a). This set of locality requirements on dependency formations follows from this path version of Minimality (henceforth, P(ath)-Minimality).

The second locality restriction is the Endocentricity Condition wherein only the head of a phrase is visible for grammatical relations. The Endocentricity Condition characterizes FL (if the proposal in Chapter 3 is correct) as only "atoms" have the power to concatenate. In particular, Chapter 3 proposes that derived concatenative atoms are the product of labeling as labeling is required if Concatenate is to apply to a previously formed concatenate. We have assumed that such labeling is (axiomatically) endocentric and that such labeling turns the complex concatenate into an atom of the type provided by the label (in accord with the understanding that labeling represents the "is-a" relation). Consequently, when an expression merges with such a labeled expression all it can "see" is the label (i.e. the head of the complex phrase) and so only this can be concatenated with (and so related to) grammatically. Thus, the system yields the Endocentricity Condition when endocentric labeling is added to the mix of basic operations.

Third, (endocentric) labeling plus concatenation result in unboundedly large hierarchically organized phrases. This arises from the fact that concatenation to complex labeled structures produces hierarchical structure. That the labeling is endocentric suffices to produce recursive hierarchy. That concatenation can

[3] That binding and control are established via movement is contentious. I believe that there is considerable evidence in its favor (see Hornstein 2001, 2003, 2006 and references therein). But the point here is not whether this is correct, but what sort of system arises if we have the modest inventory of basic operations and principles noted above. A consequence of our inventory of basic operations is a system in which construal operations are movement based. Of course, should one find the general picture attractive, then it motivates (and hence supports) treating construal as movement. One can, after all, support a particular analytical perspective both bottom up (based on empirical coverage) or top down (based on compatibility with attractive theoretical assumptions).

We have not discussed how the current set of assumptions analyzes the various "kinds" of movement, e.g. A versus A′ or restrictions like those on improper movement that regulate how they can interact. Some of these issues are discussed in Hornstein (2001). However, it is fair to say that the fine details of construal are a large and interesting open minimalist question.

apply repeatedly serves to license unbounded recursive hierarchy. Thus, a central characteristic of natural language grammars emerges from the combination of endocentric labeling and concatenation.

Fourth, labeling serves to return a labeled expression to the domain of concatenable expressions (i.e. labeling licenses further concatenation) by mapping the concatenate to one of its elements. Labeling accomplishes this by identifying the (endocentric) label with one of the concatenating lexical primes. This feature of labeling has two noteworthy consequences. First, only one of the concatenates can provide the label. For example, in A^B A might be a lexical item and B might be but AB is not. Thus, an A-labeled expression or a B-labeled one can license further concatenation but an AB-labeled one cannot. This in effect leads to phrases bearing at most one label and thus being at most single headed. Second, the labeling must conform to the strictures of Bare Phrase Structure. As X′s and XPs are not lexical items, they cannot be concatenated. Only X like things can be and so the label must be bar-free, as BPS proposes. In effect, BPS is alone compatible with the proposed labeling convention. A useful corollary of this is that BPS uniquely determines how paths are computed, in effect allowing projections of a head to count only once. The "picture" that emerges from this conception of labeling and paths is of phrases organized around heads and with little phrase internal organization of grammatical consequence. The interesting hierarchical organization is inter-phrasal not intra-phrasal. This is very reminiscent of m-command conceptions of grammatical dependencies first proposed in Aoun and Sportiche (1983).

Fifth, "bare" labeling also entails the Extension Condition. If labels are understood classically, then a concatenate labeled A "is-a" an A. As labels are all "bare" and so identical to lexical primes, this makes each labeled concatenate equivalent to a lexical prime for purposes of concatenation, i.e. concatenatively atomic. As atoms have no "insides" then the only place to concatenate is "at the edge." Consequently, Extension is the only real option. As Merge and Move are just instances of Concatenate and Copy, they must be subject to Extension.

Sixth, the combination of P-Minimality in the guise of the A-over-A condition and bare-labeling yield a version of the Structure Preservation Condition. Without further assumptions, they yield the conclusion that XPs move to XP positions, that X′s are immovable, and that head movement does not exist. As noted in Chapter 3, additional morphological assumptions can serve to neutralize the prohibition against head movement, though whether these are worth adding is currently empirically unclear.

Seventh, analyzing Merge as the combination of Concatenate and Label leaves room for an analysis of adjunction as simple concatenation without labeling. This allows for an approach to adjunction consistent with BPS.

Eighth, analyzing Move as involving Concatenate has the effect that only constituents are movable. This arises because the copy must be reintegrated into a phrase and the only way to do this is to concatenate it with the phrase. But concatenation is only licit between concatenative atoms and so the copied expression must be labeled. As all labeled expressions are constituents, only these can displace. Furthermore, if movement mediates binding and ellipsis, then these too must target constituents.

To summarize:

- The basic features of the Binding Theory and the Control module follow from the assumptions that Move is Copy and Concatenate and binding and control are products of movement, i.e. in particular, the fact that antecedents c-command their anaphoric dependents.[4]
- That Minimality involves a c-command requirement follows if relative path size is determined using Boolean measures.
- That movement is (generally[5]) to a c-commanding position follows from Move being a composite of Copy and Concatenate and from the Extension condition.
- Labels allow complex concatenates to further concatenate. This requires labels to be "bare."
- The Extension condition follows from understanding the "is-a" relation in a Bare Phrase Structure context.
- The Endocentricity Condition follows from the fact that Concatenation is the only operation for building grammatical structures (and so establishing grammatical relations) and it only applies between "bare" lexical items (i.e. heads).[6]
- The ubiquity of displacement in natural language is expected because movement (Copy and Concatenate) is required to establish relations between non-local expressions.[7]
- That labeling must return a concatenative atom prevents phrases from having more than one head.
- That Move has Concatenate as a subpart limits movement to constituents.
- If (some) ellipsis and anaphora involve movement, they will be limited to constituents.
- If labeling is endocentric (i.e. one of the concatenates supplies the label) then phrases must be hierarchical and recursive.

[4] Locality conditions on anaphora and control follow minimality and the Case Freezing Principle (i.e. the assumption that case marked DPs are no longer subject to grammatical operations).

[5] Within single rooted sub-trees.

[6] Copy does not build new structure or provide new elements to the computation. It only makes previous information available at a later derivational date. See Kobele (2006) where he notes that movement is nothing more than the process of making previously available structure available at a later derivational stage. It is the process of bringing information "forward" in the derivation.

[7] To repeat, this requires the further assumption that a single head cannot discharge all of an expression's grammatical requirements plus the assumption that not all copies are pronounced. See Chapter 7 for further discussion.

- The A-over-A condition is an instance of Minimality understood as shortest dependency as measured via paths.
- Structure Preservation Conditions follow from the A over-A instance of P-Minimality.
- That paths count each projection once (i.e. that paths are computed in terms of maxPs) follows from Bare Phrase Structure.
- Exceptions to Minimality (e.g. multiple Specs are equidistant from a common target) follow P-Minimality.
- That linearization does not permit overlapping expressions follows if linearization applies to concatenative atoms.
- If Merge is Concatenate plus Label, then adjunction structures can be treated as unlabeled, thereby allowing a theory of adjunction consistent with Bare Phrase Structure.

In sum, endocentric labeling brings in its train Extension, Bare Phrase Structure and hierarchical recursion. Combining labeling with Concatenate and Copy brings in its train c-command, displacement and the restriction of displacement operations to constituents. Restricting operations to Concatenate, Copy and Label entails movement-based analyses of binding and control with their attendant c-command and locality restrictions. If ellipsis also supervenes on movement, then all grammatical operations will necessarily target constituents. Combining this with Path-Minimality brings locality, the A-over-A condition and the minimal domain exemptions to Minimality as well as structure preservation. In short, the combination of three basic operations (i.e. Concatenate, Copy and Label) and one basic principle (P-Minimality) results in grammars with many of the features that decades of research have established to be signature properties of FL.[8] In addition, such a theory of FL provides the seeds of an answer to Darwin's Problem by laying out a possible scenario for the emergence of grammar and natural language. We turn to this next.

5.3 Basic operations and the logical problem of grammar evolution

Darwin's Problem pivots on the distance separating the grammatically from the non-grammatically endowed. If the distance is large and FL complex then large amounts of evolutionary time are required to get from there to here. If, however, the distance is small and the biological transformation needed to go

[8] The principal aspects of UG that these assumptions do not address are island phenomena. Chapter 7 suggests that some islands might be integrated into this kind of system if fully labeled paths are required for movement. Given such an assumption (reminiscent of GPSG approaches to movement) adjuncts would be impermeable for movement. However, it is currently unclear to me whether all islands could be treated as involving adjunction or if this approach is theoretically viable. See Chapter 7 for further discussion.

from non-grammatical beings to grammatical ones is relatively minor, then the short time frame available need not be much of a problem.

Given this way of conceiving things, GB style accounts appear to pose a serious problem for two reasons. First, they portray FL as having substantial internal complexity. The many modules and their complex interlocking relations are a challenge if complexity is generally the by-product of natural selection and if the time required varies directly with complexity. (i.e. the more complex the output the more time natural selection needs to work its magic). Second, many of the basic operations and principles characteristic of UG are *sui generis* in a GB like theory. For example, the Binding Theory's principles and "laws" look very specifically linguistic and so cannot just be the manifestation of more generally available cognitive resources. The more the operations and principles of FL enjoy this kind of specificity, the longer the road from pre-linguistic cognition to the emergence of grammar-based language and the greater the evolutionary time required for its emergence. The proposal outlined in Chapters 2–4 and reviewed above in 5.2 tries to solve Darwin's Problem by addressing both of the difficulties with the GB version of FL noted above.

First, in contrast to the GB picture, this account does not postulate a modular FL. In fact, the proposal is that the very same operations and conditions apply in all areas of the grammar from case and theta marking to binding and control. What differentiates case marking from reflexive binding is not the operations involved but the features manipulated. If this sort of reduction is successful it effectively eliminates the internal modularity (and much of the complexity) of FL. As noted, reduced complexity requires less time, a good result if the available time is slight.

Second, most of the postulated basic operations and principles are plausibly not unique to language. Let's consider these in turn. It is unlikely that FL is the only cognitive domain that concatenates representations or that humans are the only cognitive beings that have this particular mental operation (think action patterns or bird songs where subroutines are strung together). In short, concatenation is a very good candidate for a general cognitive primitive operation and thus its existence in and use by FL is not hard to explain. The same, I believe, can be said for copying.[9] Many animals, for example, have a small inventory of basic calls that they can use repeatedly, e.g. birds and mice string together songs from more basic "syllables" that can repeat. It is plausible that the repertoire of basic vocalization types allows for many vocalization tokens to be used again and again. If so, a Copy operation is plausibly involved. It is worth noting that both concatenation and copying are good examples of the kinds of primitive

[9] But see the Appendix in 5.7 for a discussion of a theory that substitutes multi-dominance and occurrences for classical phrase markers and Copies. The former can dispense with Copy as a basic operation. Nothing that follows hangs on which approach is adopted.

recursive operations that almost any type of computing device would use due to their general applicability. It is possible to concatenate or copy virtually any kind of representation. It is thus reasonable to suppose both that there is nothing specifically linguistic about these operations and that they cognitively antecede the emergence of FL.

What of minimality? In some form, this too is a likely part of any computational system that allows non-local relations between parts of a representation. Non-local dependencies are "expensive" to compute. They require more extensive memory resources than are demanded by local relations. As such, it is natural for cognitive representations to minimize memory load and minimizing dependency length serves this end. So, I would contend, something like minimality is a natural feature of computational systems quite generally and so not specific to language. It is in Chomsky's (2005a) terms, in part, a third factor feature.[10]

I say "in part" because Path-Minimality is a specific instance of the genus and perhaps some of its properties are specifically linguistic. P-Minimality computes distance within phrase markers in terms of paths (paths being sets of nodes) and then compares these using Boolean measures. It seems a stretch to suppose that P-Minimality is a condition available outside of FL. However, it may not be far-fetched to suppose that P-Minimality is the expression in the domain of linguistic objects of the general cognitive desideratum of minimizing memory load. This would require a way of measuring distance within linguistic objects and this is what Paths do. Actually, I find it hard to imagine any way of measuring distance between expressions within labeled hierarchical objects like phrases except via the nodes that separate them. If so, something like paths is the natural measure of distance within hierarchically organized objects like phrases.[11] Moreover, as Boolean operations are computationally very primitive, we should expect Boolean resources to be used all things being equal. If this is correct, then it is plausible that P-Minimality is the simplest implementation of the general desideratum of reducing computational load in hierarchically labeled objects. If so, this would make P-Minimality a (more or less) third factor feature of FL as well.

So it is plausible that Concatenate, Copy and P-Minimality are reflections of third factor properties of general cognition and computation in the domain of language and as such are not specifically linguistic. This leaves one further

[10] It is interesting to note that if biological memory is content addressable, as seems to be the case, we expect to find that featurally identical expressions should interfere with one another. Representations that share features are harder to distinguish accurately if memory stores representations in terms of their features, i.e. if it is content addressable. Thus, this property of minimality might reflect a general structural property of biological memory. See Lewis and Vasishth (2005) for some discussion.

[11] A similar distance measure is used in measuring "relatedness" in family trees or evolutionary clades.

operation, endocentric labeling. To my mind, this does appear to be unique to grammars. More specifically, though hierarchy is likely not unique to grammatical systems (bird songs may have something akin to syllable structure and dead reckoning systems in ants might be as well) it appears that phrasal headedness is not biologically ubiquitous. Nor is it in any clear sense a third factor feature of hierarchy. There are many serviceable hierarchical systems that are not endocentric. This suggests that this aspect of UG is a first factor property; a feature of UG that is specific to language. Let's assume that this is a biological innovation unique to FL.[12] What kind of answer to Darwin's question can we muster?

The story would go as follows: take an organism that has Concatenation and Copy, add endocentric labeling and out pops hierarchical recursion. Add non-local lexical dependencies and third factor computational considerations yield P-Minimality. With Concatenate, Copy, Label and P-Minimality many of the basic features of UG emerge. On this account, the rise of FL in humans is largely due to the emergence of a single innovation, endocentric labeling. Add endocentric labeling to the other factors, let bake for 50,000 years and out comes FL. In other words, with the right general background conditions, all that is missing for the formation of FL is one basic operation, endocentric labeling. If so, the rapid emergence of FL becomes less mysterious.

Note, this cursory story is not in itself an account of how language evolved, anymore than an answer to Plato's question is an account of how language acquisition actually operates. It (at best) provides some ingredients and points to a way of reconciling the apparent specific complexity of FL with the short time available to produce it.

5.4 Darwin's Problem and the Minimalist Program

The Minimalist Program has been motivated in various ways. In earlier work (Hornstein 2001), I adopted an epistemological perspective wherein Ockham's razor played a large role in motivating a reductive strategy towards the GB theory of UG. The argument was that reduction is always methodologically prized for if successful it broadens the empirical basis of the reducing principles. The reasoning goes as follows: if some theory, e.g. construal, can be reduced to another, e.g. movement, then this reduces the basic assumptions (axioms) required to cover the same empirical territory. This in turn increases the empirical load that each of the remaining axioms supports thereby providing each with that much more empirical support. The logic is familiar: if

[12] Nothing that follows is affected by the claim that labeling is also a third factor process. The issue would then be why humans seem unique among animals in having recursive hierarchical structure. If this is a human innovation, unique to FL then the species specificity of grammatical competence would be explicable.

four struts together support a load of 100 lbs then to support the same load each of three struts must carry 33 1/3 lbs. Take struts for axioms and load for empirical support and the virtues of reduction are transparent (assuming of course the struts *can* support the added load without collapsing). Thus is born an epistemological version of the Minimalist Program.

What the line of reasoning in 5.2 suggests is that such methodological musings can also have ontological heft. In the context of Darwin's Problem simpler theories are not merely methodologically desirable; they are, in addition, empirically valuable. There is an explanatory premium to be gained from reductive accounts because the methodological gains from reduction also address the logical problem of language evolution so long as the reduction is to theories whose basic operations have a kind of *diachronic priority*.

By "diachronic priority" I mean to evoke Chomsky's earlier concept of epistemological priority. As Chomsky observed, the primitives of UG should be such that they provide natural entry points for the learner. Notions like "subject" (and other grammatical functions) contrast with those like "agent" and "left of" in that the first is only definable inside a theory of grammar, while the latter can be defined (one hopes) independently of grammatical notions. This endows the latter notions with a kind of epistemological priority as they can leverage the learner into the grammar, which would be inaccessible otherwise. To address Plato's Problem some of the core concepts of UG must be based on concepts that have this kind of epistemological priority.

In a similar vein, theories of UG that are based on third factor features have diachronic priority in that they can support evolutionary accounts of the emergence of FL. Simply on conceptual grounds, we should expect FL to exploit operations recycled from pre-existent cognitive capacities. This is what evolution generally does. Similarly, we should also expect something additional (and idiosyncratic to FL) in the mix given the unique features natural language objects have when compared to other cognitive constructs. An explanation of FL's properties, i.e. a theory about UG, will show how the attested empirical properties of UG can be deduced from this combination of recycled and novel operations. To paraphrase Chomsky (1965), a theory about UG will be descriptively adequate if it describes the properties of FL. It will be explanatorily adequate if the uniquely linguistic features of UG when combined with the non-linguistic cognitive operations together yield the properties of FL. On the assumption that GB gives a decent first approximation of what the laws of grammar are (i.e. provides a reasonable description of FL), this translates into a research program that aims to derive GB from simpler, more natural assumptions.[13] If this is correct, then the Minimalist Program should not be

[13] See Chapter 1 for discussion of simplicity and naturalness.

understood as replacing GB but as presupposing its general validity. After all, what's the point of aiming to derive GB if one takes GB to be essentially false?

So, let's assume that an adequate minimalist theory should deduce UG's basic features as described by GB generalizations, or some rough approximation thereto. What I have argued here is that this will involve two complementary theoretical activities: (a) reducing the internal modularity of GB and (b) decomposing the basic operations of GB into complexes of simpler, more natural cognitive operations.[14] If successful these twin theoretical pursuits will provide an account of Darwin's Problem.

5.5 The Granularity Problem

Embick and Poeppel (2005b) provide a motivation for minimalism complementary to the one in 5.3. They observe that there is currently a mismatch between the inventory of grammatical primitive operations as understood by linguistic theory and neurobiologically primitive operations as understood by the brain sciences. This makes it next to impossible to link these sub-domains of research, to the detriment of each. As linguists assume that the properties of FL are ultimately tied to the neurobiological structures of the brain, it behooves linguists to start thinking about how it is that grammatical structure might be coded within brains.[15] Embick and Poeppel (2005a) concede that it will be hard to find brain correlates for the primitives of FL. This said, they propose that one aim of linguistics and neuroscience should be to solve this problem by finding a level (the right conceptual grain) that can serve to relate the basic conceptions of each. Their concrete proposal is that an appropriate level of abstraction is the "circuit."

Circuits are brain structures that compute simple operations. A useful step in the direction of bridging the granularity problem would be for grammatical accounts to "make use of computational analyses that appeal to generic computational subroutines" (Poeppel and Monahan (forthcoming)). A specific proposal is to look for basic operations plausibly dischargeable by simple and general neural circuits in terms of which the laws of grammar can be coded. This fits rather neatly with the view of the Minimalist Program outlined here. Our goal has been to find a class of very basic primitive operations that plausibly underlie linguistic computations. These same primitives are potential candidates for the primitive operations that might be reasonable building

[14] See Hornstein (2001, 2003, 2006) as well as Hicks (2006), Kayne (2002), Lidz and Idsardi (1998), and Zwart (2002) for illustrations of the reductive strategy.

[15] The opposite is, of course, also true. However, as I am here concerned to motivate a certain kind of linguistic investigation, viz. Minimalism, I will not discuss the converse issue for the brain sciences.

blocks for neural circuits. In other words, if we are lucky, the basic operations sought by minimalists will also help to solve the Granularity Problem and Darwin's Problem. All three are looking for the same kinds of basic operations.

As should be clear, neither Darwin's Problem nor the Granularity Problem is on the verge of solution. However, just as Plato's Problem served a useful function in earlier periods, Darwin's Problem and the Granularity Problem can serve to motivate research aimed at reducing the (apparent) complexity of FL and showing how it might arise from a simpler, more natural, more primitive basis. The apparent simplicity of basic operations like Concatenate, Copy and Label makes them potentially interesting, bridging primitives and thus of the right "size" for solving the Granularity Problem.

5.6 Conclusion

A FL built from Concatenate, Copy, Label and P-Minimality has many of the features of a GB version of UG. Moreover, all but Label are plausibly non-specifically linguistic operations. If so, Label is the missing ingredient required to go from an inarticulate to articulate ape.

Minimalist cognoscenti will have noticed one glaring absence from the inventory of basic operations above. Where is AGREE, the long distance feature checking operation that is central to many current minimalist analyses? This is the topic of the next chapter where I argue that this sort of long distance checking operation is neither required nor desired.

5.7 Appendix: Copies and multi-dominance[16]

The proposal outlined above assumes that Move is a composite operation comprised of Copy and Concatenate. I have mitigated my allegiance to this conception of Move by noting throughout the footnotes and in bracketed asides that a Merge/ReMerge account (reinterpreted as Concatenate/ReConcatenate) would succeed just as well for most of the issues discussed. The one place where there appears to be an argument in favor of the Copy over the ReConcatenate account involves the Extension Condition. The Copy account combined with an interpretation of labeling as returning concatenative atoms (in line with the understanding of labeling in terms of the "is-a" relation) implies that both Merge and Move are subject to Extension. This in turn underlies the derivation

[16] This section was prompted by long discussions with Tim Hunter. For a very interesting discussion comparing Copy theories with multi-dominance approaches, see Hunter (in preparation).

of the Endocentricity Condition and the conclusion that Tucking-in is not a licit grammatical operation.

Despite these (possible) virtues, however, there are some reasons against a Copy-based theory. First, there is Ockham's argument: if Move is analyzable as an instance of Merge/Concatenate alone then why add complexity by assuming Copy is involved?[17] Second, the account presented above seems to invidiously distinguish Copy and Concatenate. The latter is prohibited from applying "inside" a labeled expression because the latter is atomic. Copy, however, is not similarly restricted and "parts" of a labeled expression can be duplicated. This seems against the spirit (if not the letter) of the proposal that labeling implies atomicity.[18] Given these two considerations, it is worth reconsidering the putative problems for a copy-less account.

Before doing this, however, we must be a bit more precise about what a copy-less theory of movement requires. In such accounts, copies are replaced by occurrences. Consider an example to help fix our ideas. (1) is a representation in which α has moved from the domain of B to that of C. The dual relations into which α enters are coded in terms of the two copies of α, one in the complement of B and the other in the specifier of C.[19]

(1)

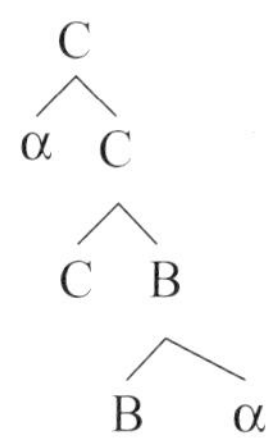

[17] It is unclear to me how strong this argument is given the perspective adopted here. If Copy exists as an available pre-linguistic operation, then assuming its availability for FL is not particularly costly. Second, if Copy is conceptually required as argued in Collins (1997) and Hornstein (2001), then once again it imposes little conceptual cost. This said, *if* Move can be analyzed without assuming Copy, then Ockham would suggest that this is the optimal approach.

[18] Again, I am not sure how accurate this is. The atomicity assumption is that labeled expressions are atomic in the sense of having no internal structure. If we take this to mean that grammatical operations cannot target the internal structures of an atom (hence, for example, lexical features are not available for syntactic manipulation nor are the "syntactic" structures inside labeled concatenates) then Concatenate and Copy can be distinguished. Concatenate must target these as it is defined as operating over defined primitives. Copy however need target nothing. It can apply freely. Of course, *if* the copy is not a concatenable then the copy cannot be integrated into the structure and so will be of little use. However, this is not a fact about Copy (which can apply freely and is not defined over primitives) but about Concatenate. In this sense, then Concatenate is subject to an atomicity requirement from which Copy is exempt, but only because Concatenate is necessarily defined to apply over specified expressions while Copy is not. This said, one might insist that "atomic" be interpreted less pharisaically as forbidding any form of grammatical manipulation, including those that are not defined over primitives.

[19] "Complement" and "specifier" are here used purely descriptively.

One can represent the same information in a phrase marker like (2).

(2)

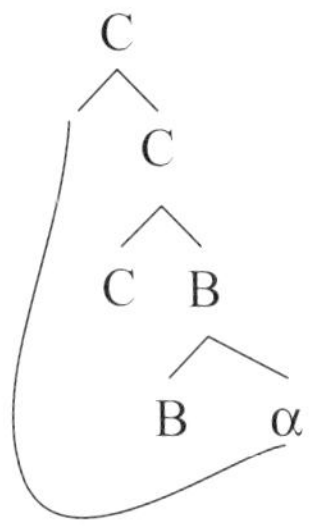

The structure in (2) codes that α is immediately dominated by both C and B (and that α is sister to both (a projection of) C and B). We can designate these the "C-occurrence" and the "B-occurrence" of α.

Structure (1) realizes the Copy theory, while (2) represents the ReConcatenate/ReMerge account. The information coded in the two phrase markers is the same. In fact, they are fully inter-translatable with copies and occurrences smoothly swapping for one another where needed.[20] The main difference is technological; where (1) uses copies, (2) uses multi-domination. Note, also, that (2) allows C to look inside a labeled concatenate to "merge" with the complement, thus apparently violating atomicity. Given this, consider the following question: How would using multi-dominance structures affect our derivation of the Extension Condition and those generalizations that we proposed are based upon it? There are several cases to consider.

Consider what happens in building (2). Early on in the derivation we have C and $[_B$ B α$]$ to concatenate. Why is the concatenation with the larger B-labeled complex (B′) and not with B itself? Consider the path between C and B′. The union of the nodes dominating the two prior to concatenation is Ø. Now consider the path between C and B. The path here is {B}. As Ø is a proper subset of {B} the latter is longer than the former and so minimality will force concatenation with B′, the large B-complex.

Will merging with B ever be possible? This depends on our particular analysis of head movement (see Chapter 3). If we assume that head movement consists of moving an X^0 to incorporate with a Y^0 then it should be possible to get a representation like (3) in which B has incorporated with a higher C.[21]

[20] See Hunter (2008) and Kobele (2006) for discussion.

[21] B will be primitive lexical atom able to meet the morphological requirements that condition incorporation. See Chapter 3 for discussion. The critical thing here is that proposal mooted in Chapter 3 using copies can be transferred wholesale to one that uses multi-domination instead. See below for further discussion of head movement.

(3)

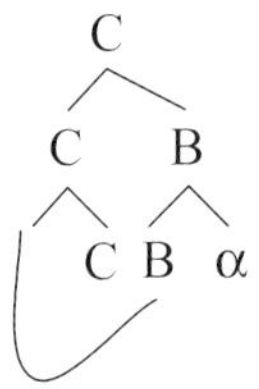

Consider now another case of movement. Assume a structure like (4).

(4)

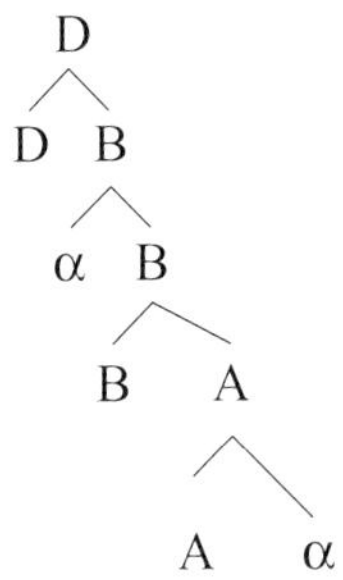

In the derivation of (4), α is copied and concatenates with B *before* D concatenates with B. The reason is that otherwise there is a violation of Extension. More particularly, once D merges with B, then a copy of α can no longer merge with B because B is inside a D-labeled expression. Given that labeled elements are atomic, concatenation with structure inside the concatenative atom is forbidden. In other words, labeling implies extension if we understand it as designating the *is-a* relation. However, what of (5) where we do not interpret labeling to imply atomicity?

(5)

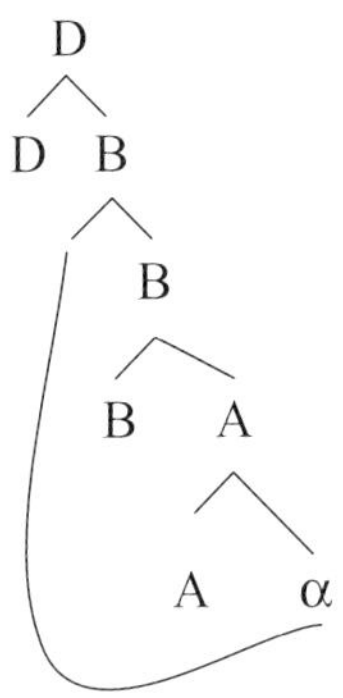

Can α ReConcatenate with B after B has merged with D? Let me be more precise: we can stipulate the Extension condition and add it to a system that allows multi-dominance structures. However, if we cannot derive Extension using independently required assumptions, this would seem to be a reason in favor of a copy-based account of Move. The question then is: Can we similarly derive Extension in a multi-dominance system? Perhaps.

Consider the path from α to B prior to the addition of D to the structure. Recall that a path is the union of the set of nodes that dominate the target and the set of nodes that dominate the launch site. In this case, it is {B, A}. Consider the path from α to B *after* D has been merged with B. It is {D, B, A}. The former is a proper subset of the latter and so the distance traversed is longer *after the addition of D*. This may seem like a counter-intuitive result. However, it seems to have the pleasant consequence of deriving the Extension Condition.

Let me be slightly more precise. If paths are defined as the union of nodes dominating the target or the launch site, then adding to the phrase marker will make any target dominated by additional structure further away from a given launch site than it was before that structure was added. If we adopt these definitions then obeying Extension amounts to taking the shortest path, which is simply our familiar path version of minimality. Or, to put this another way, there is a measurable cost to delaying establishing a relation between β and γ as soon as it is derivationally possible because any structure that gets added lengthens the "original" path from β to γ. In effect, the distance between γ (the target) and β is increased by anything that dominates γ and this has the effect of enforcing Extension if we assume that relations must span shortest paths.

This said, there are some caveats to keep in mind. First, the account requires a derivational conception of the grammar. This should come as no surprise as Extension is a notion that makes little sense in a non-derivational framework.[22] Second, the paths that are here compared occur across phrase markers. The comparison is not between two paths within a given derivation but between paths in different (albeit related) derivations. This, then, enlarges the reach of the injunction to minimize path length. Thus, if this is on the right track, minimality in its guise as Extension regulates derivational histories. It requires that relations be established via the shortest derivational routes. Third, the

[22] Interestingly, as Chametzky (p.c., 2000) observes, c-command is a representational notion. What has been attempted here has been to derive c-command restrictions from other (hopefully) more primitive notions. Extension played a large role in deriving the standard effects of c-command in Chapter 2. Here we see that Extension itself can be derived within a framework of assumptions that takes derivations as central. The upshot seems to be that notions like c-command are most comfortable in representational settings while Extension and cyclicity fit best within derivational accounts. The assumption here has been that of the two, c-command is what needs explaining and that derivations are the basis for doing so. However, whether this is correct or not, the interesting point is that c-command, Extension, and derivation all take in one another's wash and are best understood together.

result relies on a specific definition of a path. It is possible to define paths in such a way that adding structure on top of structure will not affect the distance between two expressions.[23] However, there is a palpable sense in which these definitions are more complex than the simple one adopted here (see note 23). This said, other definitions could be equally serviceable and it is unclear that Extension would be similarly derivable were they adopted.[24] All in all, however, it appears possible to "derive" Extension for these cases even without the atomicity assumption that was used to derive it in Chapter 2 if we retain the definition for *path* in Chapter 2 and give minimality a role in regulating derivational histories.

Before moving onto the next case, consider one curious implication of this proposal. Let's consider head movement one more time.

(6)

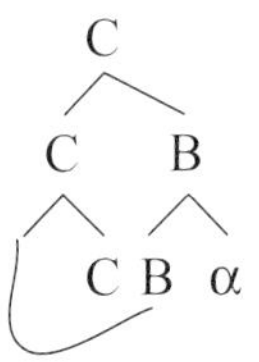

(7)

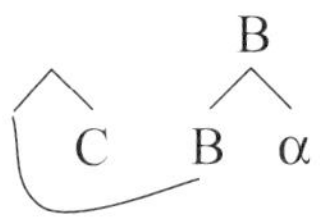

(6) represents the conventional derivation within a single rooted phrase marker where head-to-head movement violates Extension. (7) represents a derivation in which the head B sidewards moves to C, i.e. it occurs prior to the merger of C and B. Observe that the movement in (7) adheres to Extension.[25] The path account replicates the difference between the two derivations. The path in (6) is {B, C}. The one in (7) is {B}.[26] Sidewards movement has the effect of

[23] For example, if we calculated the path in (5) as the set of nodes that dominate that launch site (α) minus the set of nodes that dominate both the launch site and the target (B), i.e. the complement of the intersection of the nodes that dominate B and α.

[24] It is also unclear whether these other definitions would be empirically generalizable to all the cases discussed using the simpler definition. Thus, for example, as regards Sidewards Movement the two would appear to make different claims concerning path lengths. So, for example, movement from an adjunct to the complement of V in a *v*P structure would be longer than movement to the specifier. However, given the definition in (8) this would not be so. There are no nodes common to the adjunct and the *v*P in this case and thus the intersection would be {Ø} for both movements.

[25] See Hornstein (2001), Nunes (1995), and Uriagereka (1998).

[26] The concatenate B^C in (7) is subsequently labeled C.

shortening the path of movement. Why? Because embedding B under C in (6) lengthens the path to the head C.

Consider now one last case. As noted in Chapter 2, Extension derived via atomicity is inconsistent with Tucking-in. The logic is illustrated by considering (8).

(8)

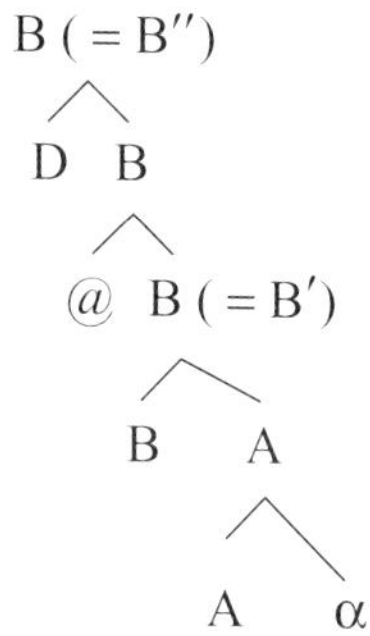

Note that the path from α to @ (sister of B′) is the same as the path to the root (B″) as all elements in the projection of a common head are equidistant from any other point in a structure. However, if Move involves Copy and we assume atomicity, then once α is copied it can only be attached to the root (B″) as the rest of the structure is invisible to it. Note too that once we copy α then the path from this copy to B″ is shorter than the one to B′. The former is {Ø} while the latter is {B}. Thus, though the distance from α to B′ is the same as that from α to BP (i.e. the chains are of the same length) the operations that go into building the structure in which α merges with B″ involve traversing shorter paths than those involved in merging with B′ once we consider the course of the derivation once the copy of α is made.[27] Thus, atomicity and minimality both forbid Tucking-in given a copy account of movement.

[27] This is simply the observation that copies, once produced, concatenate with the rest of the structure in just the same way that non-copies do. In effect, given the copy theory, Internal Merge has External Merge as a sub-operation. If this is correct, then Internal Merge cannot be an instance of External Merge. Or, more exactly, identifying Internal and External Merge requires multiple dominance rather than copies.

Note too that these observations concerning how copies are subject to P-Minimality has the effect of enforcing the extension condition without invoking the atomicity assumption proposed in Chapter 3. In other words, If Move involves Copy, then Tucking-in is impossible on P-Minimality grounds alone. This opens up the possibility of an empirical test between copy-based and multi-dominance theories.

What happens in a multiple domination account? There is no copying, recall, and no atomicity. Thus what we have is the option in (9).

(9)

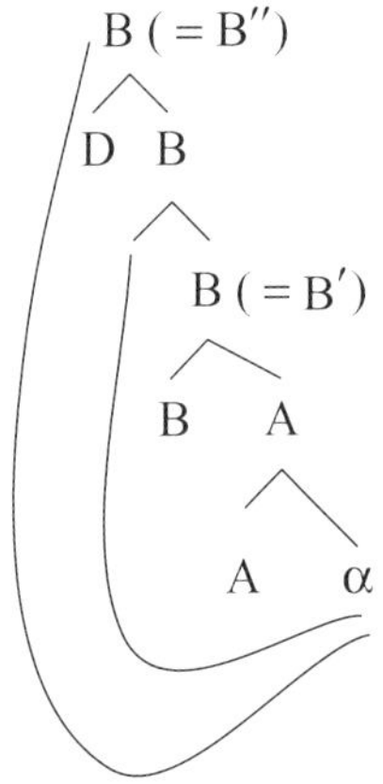

Note that the path from α to B′ is the same as that to B″, viz. {A, B}. Hence, targeting either position is fine so far as minimality is concerned. Moreover, as we have forsaken atomicity, either landing site is available as well. In short, given a multi-dominance view like the one outlined here, Tucking-in is a possibility. This still falls short of mandating it on the basis of shortest move (as in Richards 2001). However, it is not illicit, at least not on minimality grounds.[28] This sets up a nice possible empirical contrast between the Copy account and the ReConcatenate approach to Move: the former fits ill with Tucking-in and requires that apparent cases be reanalyzed. The latter is compatible with it and allows (though does not require) it.

The Copy Theory of Movement can be technically implemented in various ways. One is via a Copy operation, the other is via multiple domination structures. The aim here has been to consider whether the derivation of the Extension Condition discussed in Chapter 2 is consistent with both views. It appears that it might be (how's that for tentative!). For the larger issues discussed here pertaining to Darwin's Problem and the Granularity Mismatch Problem it does not really matter which approach to movement is adopted. What is at stake is how to understand labeling. On either account labels function to yield complex elements available for further concatenation. The question that we have been investigating is whether labels should also be interpreted as shielding their contents from concatenation. One can squeeze out this consequence given a

[28] This does not mean that it does not have other potential problems. For example, Tucking-in violates the No Tampering condition (i.e. monotonicity) and makes it harder to use the history of Merge to determine linearization. See Chapter 2 for discussion.

particular understanding of labeling in terms of the "is-a" relation. However, it appears that many of the results atomicity underwrites can be reconstrued as Path-minimality effects. Which approach, if either, is preferable is at present unclear to me. The important point here is that both are consistent with the present project.

6 Stop AGREEing! Keep Moving!

6.1 Introduction

The basic operations hypothesis limned in the earlier chapters does not include a feature checking procedure. This is an important omission, especially against the adopted minimalist background. Agreement in various forms has been a staple of minimalist analyses from the outset. Indeed, it has been a feature of *every* generative approach to grammar.[1] The reason for this is that agreement phenomena are ubiquitous within natural language. Subjects agree with predicates, antecedents with their dependents, subordinate tenses with matrix tenses, modifiers with modifiees, etc. It is hard to find a grammar of a language that does not spend considerable time on its agreement patterns. As a result, at pain of descriptive inadequacy, every theory of grammar must contain an operation that generates agreement structures.

Minimalist grammars do so as well, but with two twists. First, agreement is now an important operation and not just a widespread phenomenon.[2] In particular, since Chomsky (1993), minimalists have assumed that operations apply because they must, not because they can (e.g. as in GB).[3] This is canonized in the Principle of Greed, which requires that operations check (uninterpretable) features when they apply.[4] As agreement is the process that checks features, minimalist approaches take it to be a central operation of the grammar.[5] Second,

[1] And most likely of every non-generative approach as well given the ubiquity of agreement phenomena in natural language.

[2] To my inexpert reading of history, this is quite different from what came before. In earlier periods generative grammar thought of agreement as more a phenomenon and less an operation. This is less the case in approaches like LFG and GPSG, and HPSG where Feature Unification is a (perhaps, the) central grammatical operation. AGREE-based minimalism might, not inaccurately, be viewed as adopting the view that Unification is indeed a central grammatical process.

[3] Strictly speaking, Greed only applied to movement in early minimalism. More recently, with the proposal that Move is just an instance of Merge, all operations have been taken to be greedy.

[4] Various versions of the Greed Principle exist. Their differences are of no moment here.

[5] As movement is feature based in minimalist theories, an agreement operation is critical in the evaluation of minimality. For some discussion see Boeckx and Jeong (2004) and Starke (2001).

in more recent minimalist analyses, agreement is understood to be, in some sense, more basic than movement in that Move is a composite operation that contains agreement as a sub-part.[6] More particularly, there is an operation AGREE that operates over unbounded stretches of phrase structure in which a head α that c-commands an expression β can probe β's feature set and check "agreeing" features, thereby relating α and β.[7] If in addition to β's agreeing with α, α has some other property (e.g. EPP), then β moves to α's specifier.

Both these agreement operations go considerably beyond what is required simply to be observationally adequate. For example, agreement often applies without any morphological reflex of its application. In this sense, AGREE can be quite abstract. This kind of abstraction is not a minimalist novelty, however. It carries over from earlier GB models in which case assignment applies without any apparent morphological effects in languages like English and, more completely, Chinese. In a similar vein, minimalist analyses do not take the absence of the morphological footprints of agreement to indicate the absence of the application of AGREE.

This said, AGREE is different from the agreement operations of yore. What sets it apart from earlier conceptions of agreement (even earlier minimalist conceptions) is its application over long distances.[8] Heretofore, agreement operations only applied locally. In Chomsky (1993, 1995a), for example, it only took place within the domain of the agreeing head (e.g. Spec-head agreement being the poster child). This was also the GB view, where, for example, the long distance agreement manifested in Existential Constructions in English was taken to indicate covert movement of the associate to the specifier of the agreeing tensed head at LF.[9] On this view, the lack of locality is merely apparent; the requisite local relation being established by covert movement. Thus, in contrast to contemporary theory (in which movement presupposes agreement), previous models assumed that agreement often required movement to establish the local relation required for feature checking. This chapter argues for a return to this earlier conception on both empirical and conceptual grounds.

[6] See Chomsky (2001) and most minimalist papers since.

[7] Here on in, "AGREE" names the long distance operation and "Agree" the local feature checking operation.

[8] AGREE is roughly the combination of slash categories and Feature Unification found in GPSG and HPSG. The slash notation passes feature information unboundedly up the spine of the tree (and hence obeys the c-command condition on AGREE) and unification allows for feature checking at required points. In this sense, contemporary minimalism incorporates both the leading ideas and the technology of these grammatical approaches.

[9] See Chomsky (1986b, 2001). We revisit this kind of approach to Existential Constructions in 6.4 below.

6.2 Probing AGREE

6.2.1 The Redundancy between Move and AGREE

One of the central empirical characteristics of natural languages is the ubiquity of displacement. Displacement occurs when an expression's position in overt syntax differs from its thematic position.[10] Standard illustrations include passivization (where the logical object occurs in grammatical subject position) and Wh movement (where a thematically marked DP occurs in the specifier of a higher functional category). In typical cases of displacement, the relevant expression is phonetically expressed once and its phonological position differs from its counterpart in related, but non-dislocated, sentences (e.g. the answer to a Wh question generally occurs in a different position from the Wh element, as in: ***What** did John eat? John ate **a bagel***). To repeat, displacement is a widespread feature of natural language, one that syntacticians have tried to explain since the inception of the Minimalist Program.

Two kinds of explanations have been offered. Early Minimalism (circa Chomsky 1995a) treats displacement as an "imperfection" that only exists for its functional value. The idea is that, *ceteris paribus*, movement should be eschewed. The reason is that whereas Merge is a "virtually conceptually necessary" (VCN) operation of the grammar, Move is not. On the assumption that VCN operations are cheap, it follows that it is "costly" to move. Why then does movement apply to yield displaced structures? Because the PF or LF interfaces demand it from the grammar for their own purposes. For example, say the LF interface only computes "special" interpretations (like focus and topic) if an expression is displaced to the edge of a domain (e.g. focus accompanies movement to the clause edge in Japanese and to the *v* edge in Icelandic). The more costly movement operation is then forced upon the grammar in order to meet such LF interface requirements. Movement, then, is the price the grammar pays to synchronize the structures of grammatical objects with the interfaces that interpret them.

Late minimalism has adopted a different interpretation of displacement.[11] On this conception Move is a species of Merge. As such it too is a VCN operation. Thus, it cannot be more costly than Merge. More particularly, there exist two manifestations of Merge. External Merge is identical to Merge in prior accounts and Internal Merge is what had been called "Move." Importantly, these are *not* different operations but different applications of the very same operation. An important virtue of this view, in my opinion, is that it treats

[10] Note the abstract nature of this characterization. There are ancillary assumptions required to map overt syntax into a phonological string.

[11] See Chomsky (2005b). For earlier proposals along the same lines but based on different assumptions, see Collins (1997) and Hornstein (2001).

displacement as an expected property of natural language as displacement is the result of an operation (Merge) that is conceptually ineliminable.[12] On this conception, displacement is what a well-designed grammar produces rather than being a costly accommodation to the requirements of other cognitive systems.

It is not my intention to argue here for either of these conceptions. Rather, I would like to point out that the Late Minimalist approach fits poorly with another of its assumptions, viz. that AGREE is a more primitive operation than Move (aka Internal Merge). Here's the reasoning. In contemporary theory, Move is simply an instance of Merge, a VCN operation, and hence one that any grammar must have. Thus, a core operation of UG (Merge) is sufficient for relating non-local expressions by moving (Internally Merging) one to the other. In addition to this, contemporary theory further assumes that UG has an additional operation, AGREE, that can, like Internal Merge, relate two non-local expressions but without any displacement. This theoretical situation is conceptually odd for it embodies the following redundancy: UG equips FL with *two different* ways of establishing long distance dependencies, one via Move (itself just an instance of the VCN Merge operation) and the other via AGREE. This kind of redundancy is not conceptually optimal nor what one would expect from a "perfect" system. Moreover, if Internal Merge depends on AGREE, as is the common assumption, then the ubiquity of displacement becomes mysterious for it is possible for a grammar to discharge all of its obligations without any displacement whatsoever. In short, there exists the following conceptual conundrum: (a) why if move "comes for free" does the grammar include a second long distance checking operation like AGREE that establishes the same grammatical relations that Move suffices to establish, and (b) if UG does include AGREE and AGREE is cheaper than Move then why does displacement occur at all? It seems to me that grammars should include either AGREE or Move and displacement should be either an interface induced imperfection or a reflection of a VCN operation. The twin assumptions that Move is both perfect and costly seems a conceptually unstable mix.

Note that this reasoning only applies to AGREE the *long distance* operation. Only then is it redundant with Move/Internal Merge. An agreement operation (let's dub it "Agree") that simply effects feature checking in a local configuration is quite dissimilar from Move. The reasoning above does not argue against including an operation that allows feature agreement, e.g. Agree.

12 Collins (1997) and Hornstein (2001) propose that Move is the result of two VCN operations, Merge and Copy. This (in principle) could allow Move to be costlier than Merge and still allow it to be VCN as it is the result of two VCN operations. At any rate, here too displacement was treated as an expected by-product of the operations of an optimal grammar rather than an imperfection motivated by interface requirements.

Given the ubiquity of agreement phenomena it is unlikely that anyone would or should object to this. Rather it argues against including AGREE, the long distance agreement operation, as it is conceptually redundant with Move.

6.2.2 *Three ways to model I-agreement without AGREE*

There exist more specific theoretical reasons for rejecting a theoretical mix of Move/Internal Merge and AGREE. Current grammatical resources suffice to model I(nverse)-agreement phenomena without AGREE. Thus, adding AGREE to this mix adds further flexibility to UG. This flexibility affords UG too many options for generating an I-agreement dependency. This is both methodologically undesirable (all things equal, more brittle theories are more easily falsified, and hence better) and also creates unwanted learnability problems (all things equal, the more options UG permits the harder it is to settle on the "right" analysis). Examples will help clarify the problem.

The main empirical evidence motivating AGREE involves cases of I-agreement, a classic example of which is the Existential Construction in English illustrated in (1).[13] Note that the matrix verb appears in the singular if the embedded "associate" is in singular and plural if it is plural.

(1) There **appears/appear** to be **a mouse/mice** in the room.

Current Minimalism has the resources to duplicate the effects of AGREE in (1) without postulating a long distance feature checking operation like AGREE. Indeed, current minimalist assumptions afford *three different ways* for grammars that eschew AGREE to accommodate I-agreement phenomena. Let's consider these in turn.

One option is covert movement. This was the tack taken in the Early Minimalism (cf. Chomsky 1993, 1995a). This approach has recently fallen out of favor for it relies on multiple grammatical cycles and multiple cycles are taken to constitute a design flaw. This conclusion is debatable. It rests on the premise that AGREE can substitute for covert (LF) movement, a necessary requirement for an empirically adequate single cycle theory.[14]

How strong a consideration against covert movement accounts is this desire for a single cycle theory? It is unclear. If the cost of eliminating the LF cycle is the addition of an additional operation like AGREE, then we seem to have traded fewer cycles plus more operations for fewer operations but more cycles.

[13] Others include inverse nominative agreement in Icelandic, long distance absolutive agreement in Hindi, long distance subject-predicate agreement with inverse subjects in Spanish, and partial agreement of inverted subjects in classical Arabic. These kinds of constructions, though interesting and the subject of intense study, do not constitute the norm. Generally, agreement is rather local, which is one reason why these sorts of cases are so intriguing.

[14] See Chapter 7 for further discussion of LF movement.

Specifically, if LF movement is simply a species of Internal Merge after Spell Out, then the cost of eliminating LF seems to be the addition of an additional AGREE operation and the addition of AGREE allows for the (possible) elimination of the LF cycle.[15] So described, neither position is conceptually superior to the other. Moreover, the argument in favor of eliminating multiple cycles (and especially an LF cycle) assumes that once we eliminate covert movement for I-agreement phenomena, then there will be no further need for LF at all. In other words, it presupposes that the only reason for a covert LF component in the grammar is the need to empirically cover I-agreement. This may be so, but it is not obviously so. Many LF operations have nothing to do with agreement phenomena. Thus, it is not clear that AGREE suffices to eliminate the LF cycle and so it is not clear that the desire for a single cycle theory motivates the introduction of AGREE.

Nonetheless, for the nonce let's assume that grammars must be single cycle. There still exist two ways of modeling I-agreement without invoking AGREE. The first option involves a wider use of doubling. Sportiche (1988) launched a minor industry in syntax wherein many kinds of long distance relations were analyzed as involving movement of one part of a doubled structure. Many "long distance" dependencies have since been analyzed in these terms.[16] Sportiche-doubling is fully consistent with a single cycle theory as it relies on overt movement to break up the doubled construction. It can also be used to analyze some standard examples of long distance agreement. The next section presents an analysis of Existential Constructions (ECs) along these lines. This is of some moment for the agreement properties witnessed in ECs have long been taken to be a prime illustration of the empirical virtues of AGREE.

Furthermore, if one is inclined to be theoretically bold, then one can even imagine "pure" cases of Sportiche-doubling in which there isn't any surface hint of an overt mover. Consider what things would look like if a DP were paired with a null double and the null double moved. Say, for example, a null pronominal clitic *pro* agreed with DP (on externally merging with it) and moved to the specifier of α agreeing with it there.

(2) $[\ldots[\text{pro}_1\ \alpha^0 \ldots [t_1\ \text{DP}] \ldots]]$

On the surface this would look like an instance of long distance agreement. In sum, given the availability of Sportiche-doubling and the possibility of null

15 In prior GB analyses, the fact that LF movement was identical to overt movement was considered an important argument in its favor. See Chomsky (1981).

16 Sportiche adopted ideas first put forth in Kayne (1972). They have since been widely applied. See Aoun, Choueiri and Hornstein (2001) and Boeckx (2003a) on resumptive pronouns, Kayne (1972) and Uriagereka (1995) on clitic doubling, Kayne (2002) on construal, McCloskey (2000) on quantifier float, Rodrigues (2004) on partial control, Fujii (2006) on split control, and so on.

pronominals like *pro*, a minimalist grammar can duplicate the effects of AGREE with movement alone even given the single cycle assumption.

Current minimalist technology includes yet one more way of modeling long distance agreement without AGREE. Current copy theories of movement involve both copying and "deletion."[17] Several authors have recently explored the option of moving an expression and pronouncing a lower copy. In (2), the higher DP checks a feature of α but the lower copy is pronounced.

(3) [. . . [DP α . . .DP . . .] . . .]

Such cases would also look like cases of long distance agreement but would be mediated by overt movement rather than AGREE. Further, as the movement here is overt it is consistent with a single cycle theory. Moreover, the assumptions behind such derivations are quite conventional, given the copy theory of movement (CTM), a version of which every current minimalist account adopts. CTM has two sub-parts; a movement operation and a process that determines which copy is pronounced. The latter is often the province of case or the EPP with the copy/occurrence in the case or EPP position being the one that gets interpreted at the PF interface.[18] However, it has become clear of late that this option, though likely the unmarked case, is not the only one possible. There are times when lower copies or multiple copies are phonologically interpreted.[19] Thus, current versions of CTM combined with whatever principle makes copies pronounceable also suffice to model long distance agreement without AGREE.

The existence of three different ways of modeling long distance agreement phenomena without AGREE suggests that minimalists should be very cautious in adding AGREE to the inventory of basic UG operations. More particularly, if Move is necessarily an operation in UG (given that it is VCN) then given the various ways of modeling agreement phenomena without AGREE in a theory that simply adopts Move, the empirical arguments in favor of AGREE have to be extremely strong if it is to be admitted as a basic operation. I argue below that this threshold is unlikely to be met.

Before proceeding, however, let me outline one further reason for treating AGREE gingerly. This one is specifically related to the current project.

[17] I put "deletion" in scare quotes for it is unclear whether it constitutes a separate operation or is simply the result of not phonologically interpreting a given copy. If one assumes that anything that can be interpreted will be, then some copies might not be interpreted if they contain uninterpretable features. Under this understanding of deletion, there is no syntactic operation of deletion but just a property of the interface that prevents reading the unreadable.

[18] This way of construing movement goes back to Groat and O'Neil (1996).

[19] This idea was first proposed, to my knowledge, in Lidz and Idsardi (1998), though not quite in these terms. For contemporary versions and arguments on its behalf see Boeckx, Hornstein and Nunes (2007), Kobele (2006), Nunes (2004), Nunes and Bošković (2007) (and references therein), Polinsky and Potsdam (2006) and Stjepanović (2003).

Chapter 2 outlined a way of eliminating c-command as a grammatical primitive. In this context, AGREE is problematic. Here's why. As currently understood, AGREE is a relation between a probe and a goal and can only hold if the probe *c-commands* the goal. Thus, current conceptions of AGREE require taking c-command to be a primitive feature of UG. Chomsky (2005a,b) suggests that c-command is actually a reflex of efficient computation as it restricts the search space of the probe operation. If this is true, and I am skeptical, it conflicts with the present goal of deriving c-command rather than treating it as a primitive.

My skepticism stems from two considerations. First, it is pretty clear that restricting the search space to the sister (i.e. c-command domain) of the Probe does not in itself bound the search space (though it does restrict it). As unbounded search is computationally problematic, this restriction on Probes fails to accomplish its stated objective. In particular, Probes can look arbitrarily deep into the tree even if we assume that phases are impenetrable. The reason is that weak phases are penetrable and there is no upper bound on the number of weak phases that it is possible to string together. Thus, the search domain of the Probe and the AGREE relation it can establish with its Goal are unbounded. This severely impacts computational efficiency. In sum, restricting search to c-command domains does not in itself guarantee efficient bounded computation.

Second, there are many other conceivable ways of bounding search that would make it efficient. So the question that arises is why c-command? After all one could imagine a perfectly serviceable system that only allowed one to search one's local projection? Or one's own projection and the next one down. This would bound search even more dramatically and so, one might conjecture, would be more highly prized than the c-command condition. This is just a way of saying that the argument that c-command (or, more accurately, the condition that only allows the probing of one's sister) is not an obvious means of promoting efficient computation. If this is the point of c-command, it is not a particularly suitable construct.[20]

Let me add one last empirical-cum-theoretical point. As noted in Hornstein (2001), Probe-Goal-AGREE theories are incompatible with sideward movement. Sideward movement involves movement across unconnected sub-trees between which there is no c-command relation.[21] As I believe there to be interesting evidence in favor of sideward movement, I conclude that it is a problem

[20] I set aside here the empirical claims that AGREE is *in fact* unbounded. See Boeckx (2003a) for arguments that AGREE is unbounded and that phases and impenetrability are empirically inadequate. If this is correct, then the gain in "efficiency" by restricting search to sister of the head seems negligible.

[21] As noted in Chapter 2, Sideward Movement is not a "new" kind of movement. It is what we call the application of Move in certain configurations.

for AGREE-based accounts.[22] As noted in Chapter 2, the present account is consistent with sideward movement. Thus, *if* it exists, current AGREE-based grammars are empirically inadequate as they restrict AGREE to Probes c-commanding Goals.

This problem also extends to External Merge. If both Move (Internal Merge) and Merge (External Merge) are instances of the same primitive operation, and if Move/Internal Merge is subject to AGREE requirements, then External Merge should be as well.[23] But how can AGREE apply between two elements *before* they are merged if AGREE only applies in configurations in which the probe c-commands the goal? This casts doubt on the idea that AGREE is a precondition for the application of External Merge, and so too Internal Merge/Move if they are indeed the very same operation.[24]

This section has provided conceptual and theoretical arguments against extending UG's inventory of basic operations to include AGREE. This, to repeat, does not mean that there is no agreement operation in UG. Any account where operations must be licensed to apply (i.e. where something like Greed holds), feature checking is required.[25] The current minimalist view is that grammatical operations (including feature checking) are in service of producing structures in which all uninterpretable features are eliminated. Consequently, some form of agreement operation is *de rigeur*. However, this leaves open the question of whether this operation is local (Agree) or long distance (AGREE). The argument here has been that admitting a long distance AGREE operation into UG alongside Move should only be accepted reluctantly. Given the conceptual and theoretical disadvantages, the only good argument for doing so would be a very big empirical payoff. The next two sections turn to the alleged empirical benefits of AGREE.

[22] See Hornstein (2001) and references cited there. Of special interest is Nunes (2004). For further evidence from the Copy theory for sideward movement, see Boeckx, Hornstein and Nunes (2007).

[23] It is natural to think of selection and subcategorization in terms of feature agreement. If so, it is natural to think that External Merge is subject to AGREE no less than Internal Merge.

[24] One might consider applying AGREE after Merge rather than before. This will not have the desired effects, however. If AGREE applied after External Merge/Move then the configuration in which it applies will be very local (i.e. the domain of the head). However, in such cases after movement has applied the c-command requirement (if understood in the "no-branching category" or "sister-of" sense) will not be met as heads do not c-command their specifiers. In effect, one would have a system where if AGREE applied Move could not and if Move applied AGREE could not. This would further highlight that Move and AGREE do not mix well.

[25] One may well ask if feature checking is a distinctive linguistic operation or a property of cognitive computation more generally. It is pretty clear that it is the latter. Virtually any system that computes over the properties of a representation will need a mechanism for checking features (think edge detection features in visual perception). If this is correct, then though there may be some specifically linguistic features, checking them does not require a special kind of operation.

6.3 Existential Constructions (ECs)[26]

6.3.1 *The problem; Move and the single cycle theory*

ECs are among the most intensely studied cases of agreement without apparent movement. (4) illustrates a typical example.

(4) There **are** certain to be **mice/*a mouse** in the tub[27]

As indicated, the finite verb (*are*) and the associate (*mice/a mouse*) must agree for the structure to be acceptable. A standard analysis has the finite T^0 probe the associate thereby establishing an AGREE relation. This underlies the requisite morphological agreement in (4). The structure of (4) is (5). The matrix finite tense probes the embedded associate *mice* to check feature agreement. So goes an AGREE-based approach to ECs.

(5) [$_{TP}$ There$_1$ [T^0+are] [certain [t_1 to be [mice in the tub]]]]

However, this kind of AGREE-based account to ECs is of relatively recent vintage and it replaces earlier Move-based approaches going back to Chomsky (1986b). These analyses assume that the associate *mice* moves covertly to the vicinity of the expletive *there* (e.g. replacing *there* or adjoining to *there*). Such movement places the associate very near the finite T^0 it agrees with thus allowing the relevant (necessarily local) checking to take place. The arguments for a movement like relation holding in ECs are plentiful and compelling.[28] Here are some.

First, the relation of *there* to its associate is A-chain like. For example, the relation between *there* and *someone/a beer* in (6a,b) shows the same locality properties as between *someone/a beer* and the trace in (7a,b). This follows if the relation between expletive and associate in (6) is analogous to the A-movement relation between antecedent and trace in (7).[29]

(6) a. *There seems that someone is in the room
    b. *There is the man drinking a beer

(7) a. *Someone seems that t is in the room
    b. *A beer is the man drinking t

[26] This section is based on previously published work with Jacek Witkos, cf. Hornstein and Witkos (2003).
[27] This example is in homage to Icelandic linguistics.
[28] I say "movement like" because AGREE-based accounts are also sensitive to the locality restrictions typical of A-movement. So, in this sense, they too are "movement like."
[29] It would also follow on an AGREE-based account were AGREE (rather than Move) restricted to conditions similar to those we see in A-chains. For example, if AGREE were subject to minimality (6b) would be blocked, and if phases were impenetrable then (6a) would be blocked.

Second, there is a one-to-one correlation between expletives and associates.

(8) a. it/*there was preferred for there to be someone at home
b. it/*there was difficult for Bill for there to be someone at home
c. *There seems there to be someone in the room

If each *there* must be paired with an associate at some grammatical level (say LF), then the bi-uniqueness relation between *there*s and associates follows.[30]

Third, there is the well-known definiteness effect. Thus, cases like (9) are unacceptable. This can be accommodated if expletives must combine with their associates at LF and only indefinites can so merge. It is generally the case that definites cannot incorporate into heads while indefinites can.

(9) a. a condo for (two) girls → a (two) girl condo
b. a condo for the two girls → *a(n) the (two) girl condo

If associates merge (perhaps incorporate at LF) with their corresponding expletives at LF, this restriction would account for the definiteness effect in (10).[31]

(10) a. *There is everyone in the room
b. *There is the man drinking a lot of beer

Despite these virtues, there are several empirical and theoretical problems with a movement analysis.

First, if the associate moves then at LF it need not occupy the position that it (phonologically) occupies in overt syntax.[32] Thus, its scope position can differ from its overt position. However, this is empirically not an option. In ECs, the associate's scope is determined by its position in overt syntax. Den Dikken (1995) provides the relevant data. For example, in (11a), *many people* scopes under negation, in (11b) under the modal, in (11c) under *seems*, and (11e) does not license ACD ellipsis that is licensed in (11d).[33] In short,

[30] It is less clear how to accommodate this fact on an AGREE-based account. Typically, there is no direct relation between expletive and associate in these kinds of analyses. The bi-uniqueness requirement is thus harder to explain. One possibility would be to require *there* to check a feature against the T^0 that probes the associate or the associate itself. The technically tricky part would be to make sure that the relevant features were able to discharge these functions. For example, some of the features on *there* would have to be uninterpretable and some of the features on T^0 or the associate would have to be interpretable. I leave the technical details for others to sort out.

[31] It is not clear to me what the analogue of this would be on an AGREE account. Perhaps *there* can only probe indefinites for some reason to be determined.

[32] This would follow on an AGREE analysis if scope were not altered by Agreement. This is suggested in Lasnik (1993). However, it is not clearly correct. There are some cases of Agree that enlarge the scope of the Goal. Such cases are discussed in Ortega-Santos (2006) for Spanish and Bhatt (2005) for Hindi. Thus one cannot conclude that it is generally the case that unmoved associates scope from their overt syntax position.

[33] The ACD cases are discussed in Hornstein (1995) and Lasnik (1993).

LF movement accounts in which the associate raises to *there* wrongly predict unattested scope possibilities in ECs.

(11) a. There aren't many people in the room
b. There may be someone in the room
c. There seems to be someone in the room
d. John expects someone that I do to be in the room
e. *John expects there to be someone that I do to be in the room

Second, agreement patterns in ECs are not identical to those in which the "associate" raises overtly. ECs can display patterns of "defective" agreement while overt movement cannot. The contrast is illustrated in (12). If the morphological agreement in ECs results from the associate raising to Spec T at LF, then the contrast between (12a) vs. (12c) and (12b) vs. (12d) is unexpected.[34]

(12) a. (?)There seems to be men in the garden
b. There is a dog and a cat on the roof
c. *Men seems to be in the garden
d. *A dog and a cat is on the roof

Third, LF movement accounts are incompatible with single cycle theories. If there isn't an LF cycle then associates cannot move at LF. Compatibility with the single cycle assumption requires that the movement that relates the associate and *there* be established in overt syntax. Below, I outline an account that is compatible with the single cycle assumption and that solves the empirical problems noted above.[35]

Before doing this however, I would like to mention a problem for the AGREE-based analysis.[36] The main assumption behind the AGREE analysis of ECs like (13a) is that the associate *several books* has its case requirement discharged in the same way it is checked in (13b) (cf. Chomsky 1995a, 2000). In particular, the associate agrees with finite T^0 and thereby checks its (nominative) case. Call this the T-Agree Hypothesis (TAH).

(13) a. There were several books on the table
b. Several books were on the table

The TAH has a problem when combined with Pesetsky and Torrego's (2001) (P&T) theory of Aux-Inversion and nominative case in English. P&T makes

[34] This is also true for unmassaged versions of AGREE-based accounts. It is not clear why there should be asymmetries in agreement between AGREE alone and AGREE followed by movement.

[35] For discussion of other problems, particularly relating to the movement theory of control, see Hornstein and Witkos (2003).

[36] The following discussion borrows heavily from Hornstein (2007). For a further difficulty in the context of gerunds, see Pires (2006).

two proposals: (a) that C checks two features – [uT] and [uWH] – when T moves to C in Wh questions and (b) that nominative is the morphological reflex of [uT] on DP. These assumptions account for the presence of T-to-C in (14a) and its absence in (14b) as follows. In (14a) *what* checks [uWH] of the matrix C but being non-nominative, it cannot check [uT] of C. T moves to C to check [uT]. In (14b), *who* has both T features and Wh features in virtue of being a Wh that has moved from (nominative) subject position. Thus, it alone checks both sets of features on C. Economy blocks (14c): As *who* alone suffices to check all the features of C, movement of T is unnecessary and so unavailable.

(14) a. What did Bill see
b. Who saw Bill
c. *Who did see Bill

In light of this, consider ECs like (15).

(15) a. How many books were there on the table?
b. What was there on the table?
c. *How many books there were on the table?
d. *What there was on the table?

We derive these by moving the Wh associate to C. The contrasts between (15a,c) and (15b,d) indicate that T-to-C is required in these constructions. This follows if the associate does not have nominative (i.e. [uT]) case, *pace* the TAH. If nominative case and agreement are tightly linked (e.g. case being the reflex of agreement) then this indicates that the associate does not agree with T^0 in these cases, overt morphology notwithstanding. In other words, ECs, at least in English, are not cases of (long distance) AGREE.[37] These data are consistent with the analysis of ECs in Belletti (1988) and Lasnik (1995). These propose that the associate is case marked by *be*, not by agreement with T. There is some additional morphological evidence in English supporting this conclusion. Some cases of ECs involve definite associates:

(16) a. Who can we get to play a leading role in the spring production? There's always Bob/him/*he
b. Who can we get to play leading parts in our new production? There's always those guys in our acting class/them/*they

In the cases where the morphology is forced to appear, we find accusative case surfacing on the pronouns and nominative case being prohibited. This is

[37] Observe the same effects hold with more "long distance" cases:

(i) How many books were there believed to be on the table
*How many books there were believed to be on the table

consistent with the conclusion that in ECs finite T^0 does not mark case on the associate.

The analysis of ECs below adopts the Belletti–Lasnik proposal that in ECs the associate is not case marked by T^0. The data in (15) and (16) argue against the view that in ECs the finite T^0 probes the associate thereby setting up a long distance AGREE relation. If so, the number concord between the associate and T^0 does not arise from an AGREE relation between them.

6.3.2 *A Sportiche-doubling proposal*

The core of the proposal is that whatever chain relation exists between the associate and the expletive is due to the latter's moving away from the former overtly rather than the former's moving towards the latter covertly. This, in effect, treats ECs as an instance of doubling along the lines of Sportiche (1988). This retains the benefits of standard movement approaches to ECs and is consistent with a single cycle assumption. Consider some details.

The A-chain properties of *there*/associate pairs (illustrated in (6) and (7) above) follow straightforwardly if (17) is a case of A-movement by *there*.

(17) a. There is someone in the room
b. [There is [there someone] in the room]]

The one-to-one correlation between expletives and associates also follows on the assumption that *there* and the associate start off as a doubled constituent. The multiple merger of *theres* to associates is blocked on several grounds.

First, if *there* requires case (cf. Lasnik 1995), then stacking them would prevent them all from checking case.[38] Consider (18a). If *there* needs case, it is unclear how both instances are to discharge this requirement. One might move to Spec T to check nominative. However, there is no second case for the second *there* to discharge. Furthermore, even if there were a second case, it is plausible that the more embedded *there* cannot move across the higher one without violating minimality, i.e. if the *there*-DP in (16a) had the structure in (18b).

(18) a. T^0 is [there[there[someone]] in the room]
b. [$_{DP}$ there D^0 [$_{DP}$ there D^0 [$_{NP}$ someone]]]

[38] As Lasnik (1995) observes, the requirement that *there* bear case suffices to exclude (i):

(i) *there seems there to be someone in the room

The medial *there* is not case marked. Note that examples such as (ii) are also unacceptable:

(ii) *There is likely for there to be someone in the room

Here movement of one of the stacked expletives to the matrix position would be blocked by minimality and thus its case could not be checked.

Chomsky's (1995a) (implicit) approach to the definiteness effect would also discourage *there* stacking. It derives the definiteness effect by analyzing *there* as a kind of dummy D(eterminer) (or specifier of D). Being D-like, *there* requires a nominal (N-type) complement. If one assumes that only DPs can be definite, or alternatively that D-less NPs cannot be definite, then the fact that *there* is a D or Spec D forces the thing it merges with to be a bare NP and so indefinite. Chomsky (1995a) executes this idea by uniting the associate and expletive at LF. The present proposal differs only in having *there* merge with the associate overtly.

This can also help to account for the agreement facts in ECs. In many languages, Ds (or Spec Ds) agree with their nominal complements. For example, in French, *les*, *la*, *le*, *sa*, *son*, *ses*, *mes*, *mon*, *ton*, etc. are determiners that ϕ-agree with their nominal complements. If *there* is a dummy version of these, then it too could "agree" with its complement. This would allow the agreement witnessed in ECs to not be a function of direct agreement between T^0 and the associate, but agreement once removed. T^0 directly agrees with *there* and *there* agrees with the NP associate. By transitivity T^0 indirectly agrees with the associate.[39]

The defective agreement patterns witnessed in some ECs supports the idea that the agreement witnessed in ECs is indirect. Recall that the agreement patterns in ECs are not identical to what we find in their non-EC counterparts, e.g. we can find less than full agreement in (19a,b) but not in (19c,d).

(19) a. (?)There seems to be men in the garden
b. There is a dog and a cat on the roof
c. *Men seems to be in the garden
d. *A dog and a cat is on the roof

The defective agreement pattern in (19a,b) makes sense if the predicate directly agrees with features of *there* rather than those of *men* or *a dog and a cat*. More concretely, let's say that the English expletive *there* (in contrast to semantically robust determiners like *the*) need not agree in number with its complement. If so, when *there* agrees with finite T^0, it is a default form for number that is manifest. This is what we get in (19a,b). Note that *if* the number agreement (19a,b) is a default form (i.e. if singular agreement is what arises in the absence of agreement for number), then we expect that the converse pattern, i.e. singular associate and plural subject-predicate agreement, should be unacceptable.[40]

39 Lasnik (1995) proposes that *there* is an LF affix and that it, rather than the associate, determines verbal agreement in ECs. We have borrowed this idea (that the ϕ-features of *there* determine verbal agreement) here.

40 That singular is the default in English makes sense as this is what we find when subjects are not really specified for number:

(i) Under the table *is*/**are* quiet
(ii) It *is*/**are* under the table that I like to hide
(iii) How Bill likes to cook *is*/**are* very unclear

(20) a. *There were a man in the room
b. *There seem to be someone here

Consider another consequence of the proposal. If the associate never moves then it must occupy the position that it (phonologically) occupies in overt syntax. Thus, its scope position will be identical to its overt position. This is what Den Dikken (1995) showed to be the case as the data in (11) above illustrate.

In sum, the idea that the associate and *there* begin their derivational lives as a unit with *there* overtly (A-)moving away can explain the properties of ECs discussed in 6.3.1. In addition, it is consistent with a single cycle theory. The data noted in (15) follow on the assumption that both *there* and the associate are case marked by different heads in English, as proposed in Lasnik (1995).[41]

6.3.3 Transitive expletive constructions (TEC)

Let's now turn our attention to a cross linguistic property of ECs that has hitherto resisted a principled explanation. This involves the availability of transitive expletives across languages. In particular, they are unavailable in English, available in matrix clauses in German and in all clauses in Icelandic.[42] Consider the English data.[43]

[41] Consider the following puzzle: the expletive must precede the associate. So (ib) is unacceptable.

(i) a. There was someone here
b. *Someone was there here

This ordering follows by the A-over-A if *there* and the associate must both be case marked. The A-over-A (which recall reduces to P-Minimality) requires that the larger nominal check case before the contained nominal does. This forces the order in (ia).

[42] Lasnik (1995) provides a case-based account for this in English. We adopt part of his proposal in what follows, indicating some problems.

[43] (21d) is included to counter Chomsky's suggestions that transitive expletives do exist in English. He notes cases like (i).

(i) There were eating lunch several men that I knew

Though relatively acceptable, we take these to be due to some kind of heavy NP shift operation that takes a phonologically "heavy" NP and shifts it to the right. Note that (i) becomes unacceptable if we "lighten" the post-verbal nominal.

(ii) *There were eating lunch men

It is quite possible that the post verbal nominal in (i) is in A′ position as it seems to be best when it is on the far right periphery.

(iii) a. There were eating lunch because they were hungry several men that I knew
b. ??There were eating lunch several men that I knew because they were hungry

Moreover, it seems to license a parasitic gap about as well as more standard cases of HNPS.

(iv) a. I always recognized t right after I saw t my favorite uncle from Baltimore
b. There were t eating lunch right after I saw t several men that I knew

(21) a. *There didn't men eat lunch
b. *There didn't eat lunch men
c. There weren't men eating lunch
d. *There weren't eating lunch men

(21a,b,d) are examples of TECs in English. (21c) is *not* a TEC. Sentences analogous to (21a,b) obtain in Icelandic and in matrix clauses in Dutch and German. In what follows we first account for the absence of TECs – (21a,b) – and then say why (21c) is acceptable. We then export this proposal to German and Icelandic.

Assume that objects overtly move to Spec v. This is required in a single cycle theory without AGREE.[44] Given this, a transitive expletive construction has roughly the form in (22) if *there* moves to Spec TP overtly from the position of the associate.[45]

(22) $[_{TP}$ there T^0 $[_{vP}$ Object $[_{vP}$ $[_{DP}$ there NP$]$ *v* $[$V object$]]]]$

Note that the movement indicated in (22) is illicit. *There* has moved across the fronted object, thereby violating minimality.[46] To be more precise, though the complex of [*there* NP] and the object are in the same minimal domain in virtue of both being specifiers of the same *v* projection, *there* is not part of this domain as it is a constituent of DP. Thus, moving *there* violates minimality and the derivation is blocked. Two assumptions are required for minimality to be operative: the object must raise to Spec *v* and *there* must move from the complex *there*+associate DP to Spec TP. This, plus the definition of minimality discussed in Chapter 2, excludes TECs in English.[47]

Interestingly, transitive ECs should be permitted if the DP containing *there* moves to a position above the object in outer Spec *v*. With this in mind, consider what happens in (21c) above. Say the DP containing *there* needs to be case

[44] Lasnik (1999) provides some arguments for the assumption that movement to Spec *v*P for case is optional in English. It is natural to make it obligatory. Note that this assumption is also required to dispense with EPP features. See Castillo, Drury and Grohmann (1999) and Epstein and Seely (2006) for details.

[45] If accusative case is checked overtly, then either the verb must move around the object in overt syntax in order to get the word order right, or the lower copy gets pronounced (see 6.4). I abstract away from this short verb movement (if it obtains) in what follows. See Johnson (1991) for relevant discussion and motivation.

[46] I leave the path computations as an exercise for the reader. Incidentally, this may be a case where the two different conceptions of minimality may differ empirically. An Attract-based conception must assume that the intervening object in Spec *v* acts as an intervener despite being case marked. The more classical Rizzi account does not require this assumption. If, as noted concerning the A-over-A condition in Chapter 2, expressions that have checked their features do not act as interveners, then this favors Rizzi's original interpretation. I leave further consideration of this issue to future work.

[47] The definition in Chomsky (1995a: 356, (189), (190)) would serve equally well as it is a consequence of P-Minimality, as noted in Chapter 2.

marked/checked and this case marking/checking takes place in the Spec of *be*. Given our present assumptions, movement to Spec *be* must be overt. This yields a structure like (23).

(23) $[_{TP}$ there T^0 $[[_{DP}$ there NP] be $[_{vP}$ Object $[_{vP}$ $[_{DP}$ there NP] *v* [V object]]]]]

This derivation incorporates Lasnik's (1995) idea (following Belletti (1988)) that the associate is case marked.[48]

This derivation suggests a structure for the DP containing *there*. We can take it to be similar to genitive DPs like *John's book*. The principal difference between the postulated *there*+associate DP and genitive DPs is that *there* cannot carry genitive case, as confirmed by the fact *there* can occur in Acc-ing, but not Poss-ing gerunds.

(24) a. I would prefer there being a guard in the room
  b. There being a guard in the room annoyed me
  c. *I would prefer there's being a guard in the room
  d. *There's being a guard in the room annoyed me

If *there* cannot bear genitive case, but nonetheless must be case marked, then the only option is to move it to a case position. Moreover, given that the DP containing *there* must also be case marked, *there* must move to a position different from the one that contains the DP that it is originally a part of. This forces *there* to move *away* from its associate. In effect, this is to treat sentences like (25) as case violations.

(25) a. *[There a man] is here
  b. *I expect [there someone] to be drinking beer

This proposal further accounts for the absence of unergative ECs in English. Lasnik (1995) notes the absence of constructions like (26).

(26) *There someone jumped

Lasnik (1995) accounts for this by assuming that partitive case cannot be assigned to the associate in such structures. We follow Lasnik partway. If both the DP containing *there* and *there* need case and if unergative verbs cannot assign case in their Specs, then (26) will be a case violation; either *there* or

[48] We need not assume the case is Partitive. There is actually very little independent motivation for postulating partitive case in ECs. Lasnik (1995) uses it to derive the definiteness effect *and* to track the absence of TECs in English. It accomplishes the latter by only allowing partitively marked associates from merging with *there*. This works, but it is stipulative. The core of Lasnik's idea can be revamped and adopted along the lines indicated in the text. Vikner (1991) presents data against assuming that partitive case is involved in ECs.

the DP containing *someone* will fail to be case licensed.[49] We can repair the problem in (26) by adding a *be*, as accords both with our proposal and Lasnik's (1995) account. Here, *be* checks the DP containing *someone* and finite T^0 licenses *there*.[50]

(27) There is someone jumping

Consider one more complication.

(28) *there seems [$_{PP}$ to a man] that it is raining outside

(28) is unacceptable, though it is difficult to see why if the expletive and the associate can in principle check different cases. Lasnik (1995) uses partitive case restrictions on *there* merger to account for (28). As we have eschewed partitive case, (28) presents a problem.

An observation in Groat (1999) offers a possible solution. He notes that overt movement from within the experiencer PP with raising verbs is out.

(29) *who does it seem [$_{PP}$ to t] that it is raining

He concludes that the PP is an island impervious to movement. If so, *there* cannot move out of the PP in (28).

Let's recap. TECs are absent in English because overtly moving *there* from its DP violates minimality on the assumption that accusative case is checked in overt syntax in Spec *v*. If the *there*+associate DP overtly moves above the object, subsequent movement of *there* can occur without violating minimality.

With this in place consider German. German allows transitive existential constructions, but only in main clauses.

(30) a. Es trinkt Jemand ein Bier
There drinks someone a beer
b. *Ich glaube dass es Jemand ein Bier trinkt
I think that there someone a beer drinks

We can (almost) account for this contrast given the standard assumption that matrix clauses must be V2, viz. the finite V must be in C^0 and some XP be in

49 There is another way of deriving the absence of unergative ECs. Assume with Hale and Keyser (2002) that unergatives are actually transitives with phonetically null objects. This phonetic status can be attributed to some process similar to incorporation. What is useful here is that assumption that unergatives have objects and so are actually hidden transitives. If this object must check case, much as an overt object must, then unergatives will block *there* movement in the way that any transitive verb does.

50 It is also possible to get a kind of unaccusative example.

(i) there jumped several men

The contrast, at least for this English speaker, between presentational unergatives versus unaccusatives is not all that sharp.

Spec C. Given V2, the expletive *Es* in (30a) is in Spec C. Embedded clauses do not generally allow V to C as C^0 is filled by the complementizer *dass*, which blocks V to C (see Den Besten 1983). Thus, in (30b), *Es* is not in Spec C. Consider now the structure of (30a) if *Es* and *Jemand* first form a DP from which *Es* moves. The overt syntactic phrase marker will be as in (31).[51]

(31) $[_{CP}$ Es trinkt $[_{TP}$ $[_{DP}$ Es Jemand$]$ $[_{vP}$ ein Bier $[_{vP}$ $[_{DP}$ Es Jemand$]$ v $[_{VP}$ trinkt ein Bier$]]]$ $T^0]]$

Note that the movement of the DP containing *Es* to Spec T allows *Es* to move to Spec C without violating minimality. Note too that we must assume that the movement of *Es* to Spec C is obligatory, presumably for whatever forces V2 in matrix clauses. For English, we proposed that the expletive moves for case reasons. This seems like a less natural assumption for German given that Spec C is not generally regarded as a case position. For the nonce, assume that *Es* moves for a reason to be determined. I speculate later as to why this movement is required.[52]

With this in mind, consider the structure of (30b). There is no V2 in embedded clauses and so there is no movement to Spec C. The structure of (30b) is (32).

(32) ... $[_{CP}$ Dass $[_{TP}$ $[_{DP}$ Es Jemand$]$ $[_{vP}$ ein Bier $[_{vP}$ $[_{DP}$ Es Jemand$]$ v$[_{VP}$ trinkt ein Bier$]]]$ $T^0]]$

In (32), *Es* remains within the DP containing its associate. If, as we are assuming, *Es* must move, then this derivation crashes as *Es* is stuck within the DP. If, for example, *Es* needed case and case were assigned in Spec C, then this would crash for case reasons. However, though it is likely that the problem is

[51] We assume a somewhat older structure in which Tense is on the right side of VP. This assumption is not relevant in what follows.

[52] Note, that we are *not* assuming that the associate is marked with some secondary case provided by some additional verb, e.g. as in English with *be*. Rather the associate checks case in the standard manner in Spec T. The problem is not with the associate for the present analysis but with what forces the movement of the expletive. We mention this because transferring a Lasnik (1995) style approach to German and Icelandic (and our proposal is Lasnik-like in the relevant sense) faces two separate problems; how to case mark the expletive and how to case mark the associate. The former is a problem if the expletive does not actually go to (or through) a case position e.g. Spec C. The latter is problematic as German (and Icelandic) does not appear to have "extra" verbs like *be* around that might plausibly carry an extra case suitable for assignment to the expletive. As will become evident, we take the associate to be case marked in the standard way, i.e. by moving to a case position. This leaves the problem of the expletive discussed below.

Note too that German and Icelandic do not show defective agreement in ECs, in contrast to English. This follows if associates move to Spec T in ECs in German and Icelandic. They would thus be analogous to non-ECs in English where subjects sit in Spec T and partial agreement is not possible.

not due to case, *if* we require that *Es* move *for some reason* then the absence of movement causes (32) to crash. So, what forces *Es* to move if not case?[53]

There are some curious asymmetries between *Es* and *there* that may bear on the matter. For example, *Es* is less available than *there*, e.g. V to C can apply across *there*, but not *Es*.

(33) a. Is there someone drinking a beer?
 b. Trinkt (*es) Jemand ein Bier?

(34) a. Why is there someone drinking a beer?
 b. Warum trinkt (*es) Jemand ein Bier?

Es falls under the following generalization: expletive *Es* is only licensed in Spec C. It seems that it can only be used if required for V2. Otherwise, its use is prohibited. In this respect, *Es* functions like *do*. Let's say that *Es* is a last resort expression like *do* and that it can be used in a derivation only if licensed by some V2 requirement. This would prohibit *Es* in (33b) and (34b) as it is not in Spec C. In (30a), in contrast, *Es* is in Spec C (it is needed for V2) and so is licit. In sum, *Es* differs from *there* in not requiring case but in obligatorily occupying Spec C.

TECs exist in German matrix clauses due to the availability of Spec C as a landing site for *Es*. The DP containing *Es* avoids minimality by raising to Spec T and then *Es* moves to Spec C. This derivational two-step is blocked in embedded clauses by the absence of V2, and so the absence of an available Spec C.[54]

This reasoning extends to Icelandic. Icelandic, unlike German, is V2 even in embedded clauses. There are various ways to "describe" this. The current wisdom assumes that the subject is in Spec T in a simple embedded transitive clause.[55] However, in contrast to English and German, it is also further

53 Howard Lasnik (p.c.) suggested that maybe default case was at issue. Hanging topics in German and Icelandic can be marked with default nominative case. Perhaps, the movement of the expletive is forced to Spec C in order to get this default case marking. Interestingly, expletives in German must move to Spec C and cannot remain in Spec T. Thus, they are barred from remaining in Spec T if there is V-to-C movement (see below). This would be consistent with the idea that default case is assigned at the edge of CP, as with hanging topics. Lasnik's proposal would serve the present analysis well. I present another possibility in the text.

54 This requires the assumption that movement of *Es* to CP for V2 reasons is an instance of A-movement. Otherwise minimality won't apply. This seems reasonable for otherwise we would expect to have sentences in which the associate of *Es* is in an embedded clause while the *Es* is in the Spec of the matrix. This derivation would proceed by C-to-C movement. Thus sentences like (i) should be perfectly acceptable, but are not.

(i) *Es glaubt Johann dass Jemand im Zimmer ist
 There thinks John that someone is in the room

55 Previous analyses assumed that there was V-to-C movement in embedded clauses in Icelandic. If so, this case reduces to the German one discussed above with the additional assumption that Icelandic differs from German in requiring V-to-C in embedded clauses.

assumed that Icelandic carries an extra functional specifier position within TP. For example, Chomsky (1995a) and Bobaljik and Jonas (1996) assume that there is an extra "subject" position in Icelandic clauses. Let's assume that this is correct and let's assume, concretely, that in addition to Spec T there is another functional phrase (FP) with an available Spec. Now consider what happens in a TEC holding to the following assumptions: (a) that objects overtly move to Spec *v* to check case, (b) that the expletive starts out as a constituent with the associate and moves out of the DP that contains it, and (c) that this movement, like all movement, is subject to minimality.[56] Given these three assumptions, plus the claim that Icelandic clauses have "extra" subjects, allows the derivation of transitive expletives in both main and embedded clauses.

Consider a typical derivation, (35). The DP comprising það and the associate move over the object to Spec F. From there það moves to Spec T and the derivation converges. If Spec F is a case checking position, then the mechanics behind this Icelandic case reduces to the English examples involving *be* plus gerundive participles. But, it is worth noting that Icelandic það is similar to German *Es* in only being available where V2 is required, e.g. það is not found in questions, suggesting that it is not licensed by case.[57]

(35) $[_{TP}$ *there* V+T $[_{FP}$ [there associate] F $[_{vP}$ object $[_{vP}$ [there associate] *v* $[_{VP}$ V object]]]]]

(36) það kláruðu margar mýs ostinn alveg
There finished many mice the.cheese completely

If so, the relevant licensing condition for the Icelandic and German expletives is that they end up in Spec C to license V2.[58] English *there*, in contrast, must be case marked. The assumption that expletives form units with their associates in overt syntax and then move to be licensed plus independently motivated assumptions about V2 structures in German and Icelandic suffice to account for the distribution of TECs in the three languages.

6.3.4 *Recap*

If we assume that AGREE is not an operation of the grammar then ECs must be products of movement. Combined with the assumption that grammars are single cycled prevents this movement from being covert. This eliminates the

[56] We first consider cases where the object seems to overtly move and then consider cases where this movement need not occur. Note that if objects do not move to Spec *v* then nothing should block the generation of transitive expletives.

[57] There is further discussion of the Icelandic facts in Hornstein and Witkos (2003) concerning apparent violations of minimality. The interested reader is referred to the discussion there.

[58] This holds for those expletives analogous to *there*. These are the ones that show inverse agreement and must appear in Spec C. The ones analogous to *it* do not show inverse agreement and need not appear in C.

possibility that in ECs the expletive and the associate combine at LF. However, these assumptions are compatible with a Sportiche-doubling analysis in which the expletive and associate first combine and the expletive separates by (overt) movement. I have rehearsed the empirical virtues of this kind of analysis and shown how it can handle the core properties of ECs and can extend to offer an account of TECs cross-linguistically. I know of no empirically comparable AGREE-based account. If so, this argues *against* an AGREE-based analysis of ECs. However, even if an AGREE-based account exists with equal coverage the real interest of this analysis is that it eliminates one important category of empirical support for AGREE-based systems if viable. Combined with the AGREE system's other problems it reinforces our conclusion in 6.3. that AGREE should be eliminated as a UG operation.[59]

6.4 Pronouncing lower copies

Not all cases of I-agreement display the properties we find in ECs. For example, there are cases of I-agreement that do not display a definiteness effect or in which partial agreement is not an option or in which a DP's semantic scope is wider than its overt position would license. For these sorts of cases, the doubling logic outlined in 6.3 is inappropriate. These kinds of constructions have been investigated by a variety of authors in a variety of languages.[60] In contrast to 6.3, my aim here is not to provide alternative analyses of these constructions but to show that a movement account can do as well as an AGREE-based one *regardless of the data involved*. In short, the issue is not actually one that can be empirically adjudicated as grammars that have movement and allow lower copies to be phonologically interpreted will be empirically indistinguishable from those that contain AGREE.[61] The reason is that AGREE is just a special

[59] We noted above that adding AGREE to a standard minimalist theory makes it suppler, and that this is a bad thing. The more supple a theory the more analytical options it allows. This both lessens its empirical support and, understood as a theory of UG, has adverse impacts on learnability issues as it widens the class of possible analyses of roughly the same stretch of linguistic data. In other words, the mechanisms will cover large swaths of the same empirical territory and this is bad from both a learnability and methodological point of view. Note that even eliminating AGREE leaves us with more than one grammatical option: covert movement, Sportiche-doubling and lower copy pronunciation. One argument in favor of a single cycle theory assumption is that it would eliminate covert movement as a possible option. If UG limited Sportiche-doubling to those cases where there is an overt double (e.g. *there*) or where an easily observed signature property signaled its application (e.g. the definiteness effect) then the choice between the two remaining options would be greatly simplified. Whether this is achievable, however, is not something that I can settle here. I leave it for future research.

[60] The languages include, among others, Hindi-Urdu (Bhatt 2005, Boeckx 2004, Chandra 2007), Spanish (Ortega-Santos 2006), Tsez (Polinsky and Potsdam 2006) and Classical Arabic (Soltan 2007). This is just a small selection of the relevant literature. For a fuller bibliography, see Chandra (2007).

[61] To be entirely accurate, standard data of the sort exploited by grammarians will not settle the issue. See Chapter 7 for some discussion of the "psychological reality" of compiled operations.

case of movement, one in which the lower copy is interpreted. As Kobele (2006: 143) has aptly phrased matters: "[A]gree as a grammatical operation is simply movement with particular interface effects." As virtually every current minimalist theory includes both the Copy Theory of Movement and the option of pronouncing lower copies, AGREE accounts cannot empirically cover cases that a movement account cannot. In short, wherever an AGREE-based analysis is used to explain the properties of some construction, it is possible to mimic the account by supposing that movement has occurred, the higher copy has been deleted and the lower one retained. Thus, as a point of logic, the empirical coverage of AGREE-based accounts cannot be superior to Move-based ones. Let me illustrate this logic with an example from Icelandic.

Consider a standard case of I-agreement in Icelandic. Icelandic sentences with quirky case marked subjects allow another DP to agree with the finite predicate and get nominative case. Abstracting away from possible movement to CP for V2 reasons, the structure of (37a) is (37b).

(37) a. Henni voru gefnar bækurnur
She.dat were.pl given.pl books.nom.pl

b. $[_{TP}$ She$_1$ $[[_T$ past+were$]$ $[_{vP}$ t$_1$ v $[_{VP}$ given books$]]]]$

An AGREE-based account generates (37b) as follows: after *she* raises to Spec T, T probes its complement domain to check its uninterpretable ϕ-features.[62] *Books* is T's goal as it has the relevant interpretable ϕ-features. An AGREE relation is thus established between T and *books* and T's features are checked and valued. The nominative case on *books* is a reflex of this AGREE relation.[63]

A Move account could assign (37a) the structure in (38).

(38) $[_{TP}$ She [books $[_T$ past+were$]$ $[_{part}$ books $[_{vP}$ she v $[_{VP}$ given books$]]]]]$

The agreement on *given* and *were* arise as *books* moves into the Spec of the participle head and then the Spec of T. In these constructions, the lower copy of the chain is phonologically interpreted.[64]

These analyses are virtually isomorphic. Where in an AGREE-based system only AGREE applies, in a Move-based system Move applies and the bottom

[62] The probing must follow the raising of *she* for otherwise *she* would block this agreement (an instance of defective intervention (aka minimality)). This violates strict cyclicity as the probe does not check its features as soon as it can. One can relax cyclicity in various ways, e.g. cycle only on the phase so that strict cyclicity does not hold within a phase, cycle on maximal projections to the same effect. The latter is problematic, however, as given BPS an unembedded node is always maximal. One would need a notion like a "closed off" node rather than a maximal one. Thus a TP with a filled Spec is "closed off." Observe that this succeeds by adding another primitive relation to the grammar, viz. α closes off β.

[63] The agreement on *given* arises similarly with *given* acting as the probe and *books* as goal. There is no case assigned, however, as *given*'s ϕ set is incomplete. It does not contain a person feature.

[64] The participial head might just be *v* with participial features. One further point: As noted above, it is a standard assumption concerning Icelandic that it has multiple subject positions. Whether this is coded as allowing multiple specifiers of T or another FP with a specifier (as above in (35)) does not matter here.

copy is pronounced. Where in an AGREE-based system AGREE plus Move apply, in a Move-based system the top copy is pronounced. Every AGREE-based analysis can be mapped into one involving movement. Here's the recipe: in cases of I-agreement assume that the "goal" has moved to a specifier of the "probe" in overt syntax and that the lower copy of the goal is pronounced (rather than the higher copy in the specifier position of the Probe).

One might object that such a Move account offers no explanation for *why* the bottom copy is pronounced on some occasions while the top one is pronounced on others. This is correct.[65] But in this it does not differ from AGREE-based accounts which suffer an analogous (identical?) defect, viz. why in some cases of agreement are there EPP requirements forcing movement to the Spec of the probe while in other cases there aren't? The only real difference between the two approaches is that whereas an AGREE account still requires Move, a Move account can do without AGREE. In this sense, AGREE is a special case of Move and, as such, it is both conceptually superfluous and without empirical advantage.[66]

In fact, AGREE may introduce unwanted complications. I-agreement in Icelandic has binding consequences, illustrated in (39).[67]

(39) a. *?Konunum$_1$ fundust pær$_1$ vera gáfapar
women.the.dat seemed.**3pl** they.nom be gifted.fem.pl.nom

b. Konunum$_1$ fannst paer$_1$ vera gáfapar
women.the.dat seemed.**3sg** they.nom be gifted.fem.pl.nom
"It seemed to the women they were gifted"

65 There are many good reasons for why, in particular cases, lower copies must be pronounced, e.g. Bošković (2002), Stjepanović (2003). Polinsky and Potsdam (2006) and Boeckx, Hornstein and Nunes (2007) also offer some speculation. However, at present what drives high versus low copy pronunciation, if there is indeed anything general that drives it, is unclear. Note that so long as *which* copy is being pronounced is easy to cull from the primary linguistic data, there need not be a general account.

66 This discussion assumes that agreement is with T in Icelandic I-agreement. This may be incorrect. The nominative DP in (37) and (39) might not be agreeing with T but with some other functional projection. There is a curious third person restriction on this kind of I-agreement. Thus sentences like (i) are unacceptable with non-third person nominative I-agreeing DPs.

(i) *Henni leiddumst vid
Her.dat bored.3pl us.nom
"She bored us"

This restriction is odd as Icelandic allows first and second person subjects. This suggests that the agreement seen in Icelandic I-agreement structures is not provided by T. See Boeckx (2003b) for discussion. For present concerns whether this is correct or not is not important. The relevant point is the translatability of AGREE accounts into Move accounts without empirical leakage.

67 Data from Taraldsen (1996). Ortega-Santos (2006: 58) cites Spanish binding data noted by Zubizarreta that leads to a similar conclusion. The paper observes that it is possible to bind pronouns with quantificational antecedents even in the apparent absence of surface c-command and that this can be accounted for if one assumes that the copy that is pronounced is not the one that is the binder.

The relevant difference between the two sentences is that in (39a) the matrix verb agrees with the embedded subject while this is not so in (39b), where the matrix has default third person form. If we assume that agreement requires movement, then there is a copy of *they* in the matrix clause in (39a) but not in (39b). Thus, *the women* and *they* are clause-mates in (39a) but not (39b). It is thus not surprising that Principle B effects appear in the former but not the latter. In effect, agreement on a higher predicate alters the scope of the agreed-with expression, moving it higher up the tree. This is not what we expect on a pure AGREE-based system. In fact, this sort of scope alteration is counter to what exists with ECs as noted in the discussion of the Den Dikken (1995) data in (11) above.[68]

In sum, the maneuver outlined above demonstrates that excluding AGREE will not compromise the empirical coverage of Move-based grammars so long as the option of pronouncing lower copies is permitted.[69] As this latter option is independently required if the Copy Theory of Movement is adopted, deleting AGREE from the inventory of basic operations cannot compromise data coverage.[70]

The two approaches may, however, differ in explanatory power. Here's what I have in mind. Let's assume a classical conception of Greed, one in which movement occurs to check features of the moving element.[71] It embodies the idea that grammatical operations are in the service of discharging uninterpretable features, which if not discharged will lead to an interface crash. If this is correct, then Greed licenses grammatical operations by requiring that they forward the process of feature checking.

In this light consider the Copy Theory of Movement under an interpretation in which Move is Copy plus Concatenate/Merge rather than Internal Merge.

[68] One can make AGREE-based accounts consistent with these data by assuming that binding domains follow agreement domains. This is essentially proposed in Bobaljik and Wurmbrand (2005). See their paper for other cases where agreement has scope effects. See in addition Bhatt (2005) for similar effects in Hindi-Urdu and Chandra (2007). What the above suggests is that this adjustment to the binding theory is unnecessary once AGREE is eliminated in favor of Move.

[69] See Nunes and Bošković (2007) for a thorough review of cases where lower copies are pronounced. As their survey makes clear, the assumption that one can have lower copy pronunciation is empirically quite well established and is part of most every minimalist proposal. See Polinsky and Potsdam (2006) for additional material.

[70] This claim assumes that AGREE duplicates the locality conditions affecting Move, e.g. minimality. This is the standard assumption. Effective intervention covers virtually the same territory as minimality. However, there have been proposals that subject AGREE and Move to different restrictions (e.g. Bošković in press). The argument here assumes that the standard view is the correct one.

[71] I believe that the argument will go through, though not as neatly, if we assume that at least some items of the mover are checked and that a DP stops moving when all of its features are checked. This is roughly the condition that says that a case marked/checked DP is not subject to further (A-)movement.

Lower copies will typically carry more uninterpretable features than higher copies (typically case or some A′-feature). This is why these DPs move. Each "movement" serves to check a feature of the DP. More exactly, higher copies will typically be less encumbered with uninterpretable features than lower ones. For example, a case marked DP will be higher than its copy that is not case marked. If we assume that the copy that is pronounced is the one stripped of uninterpretable features (e.g. with checked case), then typically (i.e. in the unmarked case), the highest copy will be pronounced. This is the insight captured in the old Chain Condition, which required that the lowest link be in a theta position and the highest in a case position in well-formed chains. On this view, then, the unmarked situation will be the one in which the highest copy is pronounced. The marked case will be the one in which the lower copy is.[72] Now, as a matter of fact, examples of I-agreement display the signature features of markedness. They are far less common than cases of Spec-head agreement and are often far more idiosyncratic in their agreement patterns (i.e. they show defective agreement).[73] They may also be restricted to a handful of contexts. For example, Pritha Chandra (p.c.) informs me that the cases of I-agreement in Hindi-Urdu discussed in the literature are restricted to two verbs. This is what one expects from a marked option. So, if I-agreement is the product of Move plus delete-the-higher copy we have the beginnings of an account as to why it is less common than Spec-head agreement, why it often displays incomplete agreement patterns and why it is often lexically restricted. It is because it is the marked "abnormal" case.[74]

This is not so on an AGREE-based account. Here the simpler derivation is one in which AGREE applies without movement. Movement is an add-on and does not serve to check any features at all. It is driven by an EPP requirement, which is expressly conceived as *not* checking any features of the mover. On this view, movement is quite unnecessary for the purposes of feature checking and convergence. It is there to meet EPP requirements. As such, insofar as there are any expectations, I-agreement should be the unmarked case as it is computationally simpler and movement to the Spec of one's probe should be the marked one. In effect, the two conceptions lead to opposite expectations. As noted above, it is my impression that cases of I-agreement are less common and more idiosyncratic than cases of Spec-head agreement.[75] If this is so it

[72] On the assumption that the higher copy will typically be the case marked one. Note, that the less encumbered copy is the one targeted for pronunciation all things being equal does not mean that things are always equal.

[73] For example, in classical Arabic I-agreement codes gender but not number, and in Icelandic it is restricted to third person DPs.

[74] One might conjecture that the defective agreement "signals" that the lower copy gets pronounced.

[75] This is an empirical issue so "my impression," given that it is not backed by much, should be treated gingerly. At any rate, these two approaches lead to different conceptions of what the "standard" case should be.

empirically favors a Move-based conception of how long distance dependencies are discharged.

6.5 Conclusion

Most contemporary minimalist theories of grammar assume that AGREE is part of UG's inventory of basic operations. This chapter has argued for a reconsideration of this assumption. The argument has three strands.

First, AGREE should be rejected on methodological grounds as it introduces redundancy into the grammar. It does so in two ways: (a) if Move is actually an instance of Merge and Merge is virtually conceptually necessary, then UG has the wherewithal for coding long distance dependencies without AGREE.[76] Hence, *ceteris paribus*, there should not exist a second way of executing such dependencies that relies on different operations, especially if these are not similarly VCN, e.g. AGREE. (b) The main empirical support for AGREE involves cases of I-agreement. However, even without AGREE, standard minimalist accounts have operations sufficient to code I-agreement. In fact, there are three different ways; via covert movement, via Sportiche-doubling and via the phonetic expression of lower copies. These options exploit standard technology and so adding AGREE as a separate operation introduces redundancy into the theory. This is methodologically undesirable and adversely impacts learnability.

Second, within the context of the present project AGREE has the unfortunate property of being defined in terms of c-command in that Probes must c-command Goals. As part of the present project is to remove c-command as a grammatically primitive relation and deduce it from independent factors, adding AGREE to UG is undesirable.

Third, the empirical evidence adduced to date in favor of AGREE is not sufficiently compelling to weaken the theory by adding it. Section 6.3 provided an alternative analysis of ECs that is both compatible with the single cycle hypothesis and that is empirically well grounded. Indeed, in my view, it is superior to most standard accounts in providing an account of the distribution of TECs. Section 6.4 argues that given the Copy Theory of Movement and the need to decide which copy to pronounce, minimalist theories already contain the option of deleting lower copies. This option is empirically well grounded and it suffices to model all cases of I-agreement without AGREE. Indeed there is a straightforward recipe for translating any AGREE-based account into a Move-with-lower-copy-interpretation story. If this is correct, there *cannot* exist empirical evidence of the conventional kind for choosing between a minimalist

[76] Similarly, if Copy and Merge are VCN (as discussed in Chapter 5).

grammar with AGREE and one without, so long as it includes the Copy Theory of Movement.[77]

There is one last reason for dispensing with AGREE; it makes the ubiquity of displacement mysterious. If derivations involve feature checking(/assignment) and such checking is only licensed in very local environments (e.g. the projection of the checking head) then given that expressions must check multiple kinds of features it is not surprising that natural language grammars show a lot of displacement. Movement is the only way of meeting these multiple feature demands. In an AGREE-based grammar, movement is entirely adventitious. Its source is interface requirements. But for these, everything would stay put. This may be correct, but then displacement is not really an inherent property of grammars but the functional result of the fact that grammars interface with other cognitive components that place demands on its products. Given that most of what we "know" about the properties of the interfaces are little more than stipulations, this is an unsatisfying kind of explanation to the question of why grammars so often displace expressions. It is heartening to observe that current conceptions of Move as Internal Merge endorse this dissatisfaction. To allay it, however, we should remove AGREE from the inventory of basic UG operations.

[77] Note that between the Sportiche-doubling account and the Pronounce-Lower Copy account we have the wherewithal to account for cases in which the I-agreeing expression can (e.g. Spanish, Hindi, etc.) and cannot (e.g. English ECs) scope higher than its apparent overt position. As both kinds of cases are attested, it suggests that both kinds of options are required. It raises the important question of how children learn which apparatus underlies any given case of I-agreement.

7 Conclusions, consequences and more questions

7.1 What we have wrought

In the previous six chapters I have tried to outline a minimalist project that takes the "success" of GB (and kindred theories) as starting point.[1] Thus, I have assumed that GB has "more or less" correctly identified (many of) the "laws of grammar," e.g. the "law" that reflexives must be locally bound, that binding necessarily involves c-command, that minimality governs movement, that only c-commanding interveners count in determining minimality, that movement is (in the standard cases) to c-commanding positions, etc. I then proposed that we consider GB an "effective" theory in the sense that we treat it as roughly empirically correct and try to derive its "laws" from more general principles.[2] The word "roughly" is important here. For example, there may be cases in which a GB claim is empirically qualified. A case in point: as noted in Chapter 2, there may well be cases of binding without c-command and such cases could prove decisive in evaluating the enterprise. However, the project outlined here (and the Minimalist Program quite generally) starts from the assumption that the empirical generalizations at the core of GB and its generative cousins correctly describe the lay of the grammatical land. The novelty of minimalism (and the present project) is to take GB's success as posing a *theoretical* challenge; to solidify these results by grounding them in deeper and more natural principles.[3]

The main motivation for the theoretical project is provided by the evident complexity of GB style theories.[4] In particular, FL, on the GB conception, is both internally modular and replete with principles and primitives that are very language specific. These features of GB raise two problems. The first,

[1] From here on out I will only mention GB. However, the reader should hear echoing in his/her ears "as well as GPSG, HPSG, LFG, etc."

[2] This sort of project was undertaken once before when Chomsky (1977) proposed deriving Ross's island conditions from the more general notions of subjacency.

[3] This is exactly parallel to Chomsky's (1977) project with respect to Ross's Island Conditions. There too it was assumed that Ross's description was roughly empirically correct. The aim was not to displace Ross but to ground his discovered generalizations in more fundamental computational principles.

[4] See Epstein (1999) on some of the intricacies of GB.

Darwin's Problem (or the logical problem of language evolution), rests on three assumptions: (a) that complexity is the product of natural selection, (b) that the more complex the product the more time natural selection needs to operate, and (c) FL has emerged too rapidly for natural selection to work its magic, the time frame being roughly 50,000–100,000 years. If this is correct, then the only reasonable conclusion is that FL cannot be as complex or as exceptional in its principles and basic operations as GB makes out. Or to put this positively, UG is less internally modular and its basic operations and relations are more generic than GB suggests.

The second problem concerns the realization of UG in brains. David Poeppel and friends have forcefully argued that the basic constructs and concepts of the brain sciences fit poorly with those of linguistics. There is, in his words, a "granularity mismatch" between the two, which makes it hard to see how the operations invoked by UG could be realized in what neuroscientists think of as neural circuitry. If we assume that FL is embodied in the brain, then it is worthwhile exploring whether the basic concepts of linguistics might be cast in terms that are more amenable to wet-ware embodiment. A useful step in this direction would be for grammatical accounts to "make use of computational analyses that appeal to generic computational subroutines" (Poeppel and Monahan (forthcoming)). A specific proposal is to look for basic operations plausibly dischargeable by simple and general neural circuits in terms of which the laws of grammar can be coded.[5] Note that an appreciation of Darwin's Problem would encourage a similar kind of hunt. To the degree that FL exploits generic neural circuitry similar to what is used in other parts of the brain, the smaller the evolutionary distance that must be traversed and the easier it is to account for the rapid emergence of FL. Thus, both the "logical problem of language evolution" and the "Granularity Mismatch Problem" call for grounding the laws of grammar (as outlined by GB) on a simpler and more general foundation.[6]

If this is correct, the right theory of grammar will be one that has (roughly) the empirical coverage of GB, and that "solves" Plato's Problem, Darwin's Problem

[5] Lest there be any misunderstanding, this is *speculative* neuroscience. Poeppel's proposal for solving the Granularity Mismatch Problem is to look for simple general operations that underlie the apparent grammatical complexity. The most the present proposals can lay claim to is that the suggested basic operations are of *the right kind*. Nothing more specific can be claimed. To go beyond this, real, not speculative, neuroscience is required. And, in case this is not evident, let me note that none has been provided here.

[6] Note that these reasons are different from the methodological motivations that the minimalistically inclined might have for pursuing a simplification of GB. Hornstein (1995) tried to construct a methodological rationale for the Minimalist Program. Whatever one thinks of these, the two adumbrated here are quite different. They are broadly empirical considerations, like Plato's Problem earlier on, rather than methodological considerations like Ockham's Razor. Of course, both kinds of reasons are worthy of consideration and can even be mutually supporting. My only point here is that they are different in kind and are worth keeping separate.

and the Granularity Mismatch Problem. The aim of the earlier chapters has been an immodest one; to roughly outline one such possible theory. The approach has been reductive in two ways.

First, I have assumed that it is possible to eliminate (or at least significantly reduce) the internal modularity of FL by reducing the disparate operations of GB to a common core that includes Merge and Move (itself a species of Merge). Large parts of early minimalist work can be understood as showing how to reduce case and agreement to movement. The work (not reviewed here) that reanalyzes construal in terms of movement, e.g. the Movement Theory of Control and movement treatments of anaphoric binding, helps further reduce the internal modularity of FL and reduces the operations required to establish grammatical dependencies. In the best of all possible worlds, the various GB modules are a simple reflection of the same Merge and Move operations applied to different kinds of primitive items (i.e. lexical items and their features). If correct, FL has no internal modular structure.

This book has focused on a second reductive strategy. It has outlined one way to deduce a portion of the GB theory of UG (suitably modified so as to be non-modular) by assuming that FL uses three basic operations – Concatenate, Copy and Label – and that these operations apply in a way consistent with minimality (construed in path terms).[7] The interaction of these operations under this condition yields structures that look a lot like those we find in natural language. From it we get hierarchy, cyclicity, c-command, constituency, equi-distance, locality, etc. as discussed in Chapter 5. Moreover, of these basic operations and principles only Label clearly emerges as specific to language. The others are plausibly either operations that apply in other cognitive domains (viz. Concatenate and Copy) or are reflections of general principles whose function is to minimize the computational load that non-local dependency imposes (viz. minimality). If this is correct, we have a candidate answer for Darwin's Problem: evolution of FL can be rapid because there is really only one basic operation we need to add (viz. Label) to the inventory of previously available cognitive operations and principles in order for brains to have the wherewithal to generate linguistically shaped expressions. We also have a remedy for granularity mismatch: there are three basic circuits: Concatenate, Copy, Label. All three are simple enough for neural structure to embody. Two (viz. Concatenate and Copy) are generic and likely operative in other parts of our cognitive economy. One (viz. Label) is proprietary to the linguistic domain. It is the linguistically quintessential operation. In sum, the proposals in

[7] Two points: There is a fourth operation, Agree, that is also part of the package. However, as we have said relatively little about how this operation functions (except to argue that it is local and likely not distinctively linguistic) I will set it aside here. Second, recall the discussion in Chapter 5 where we suggested that Copy might not be part of the FL mix either. What follows is neutral with respect to the question of whether Copy is a basic operation.

Chapters 1 through 6 provide a model for the kind of theory we should be looking for; one that is reasonably well empirically endowed and that provides possible and plausible answers to the three kinds of questions that a fully adequate theory of natural language should address.[8]

7.2 Some consequences

It strains credulity (even my own) to believe that the current proposal is correct, at least in detail. However, for the remainder of this chapter I want to assume (no doubt incorrectly) that the kind of approach outlined above is roughly adequate and consider some possible consequences for the study of FL.

7.2.1 Complex operations and constructions

First, in terms of the present account, at the most fundamental levels, FL does not contain operations like Merge and Move. On the present account, both result from the serial application of several operations, Merge being Concatenate followed by Label, and Move being Copy followed by Concatenate, followed by Label. Nonetheless, one can ask whether Merge and Move exist as *composite* operations, if not in UG, then as part of a person's particular grammar. Here's what I mean.

It is possible that native speakers compile primitive operations and that the results are complex operations like Merge or Move. A compiled operation could then be treated as a primitive within a native speaker's grammar even though the compiled operation is not itself a primitive of UG, i.e. not a part of FL that is genetically inherited. In fact, there are some minimalist proposals that presuppose this. Let me illustrate.

Consider the economy condition Merge-Over-Move (MOM). Chomsky (1995a) proposed that this applies in cases like (1a) to prevent the derivation of (1b).

(1) a. [to be a man here]/ Numeration: {there, seems}
   b. *There seems a man to be here

The idea is the following: at the point of the derivation depicted in (1a), it is more economical to Merge *there* from the numeration than it is to Move *a man* in the derivation. One motivation for this reasoning is that Merge applies in preference to Move because it is a cheaper operation than Move. How so? Because Move contains Merge as a proper subpart given that Move = Copy plus Merge. This

[8] I want to emphasize, in case it is not obvious, that these are *possible* answers. We are dealing with the *logical* problem in each case. We do not know, for example, if brain circuits really are concatenative. However, this is the kind of circuit that is reasonable to expect in being general, simple, ubiquitous, and embodieable.

reasoning presupposes that Move is a compiled operation (i.e. that Move is more complex than Merge) and that simpler operations apply before more complex ones do. Speaking economically, it is *cheaper* to apply Merge than Move as it involves a proper subset of the operations in Move.

This argument only goes through as stated by assuming that Move is a compiled operation. To see this, assume that in place of Move we have the serial application of Copy and Merge. At the point of the derivation depicted in (1a) we have two options; either copy *a man* or merge *there*. We can apply either operation *at this point of the derivation*. On the assumption that *ceteris paribus* no primitive operation is inherently more costly than any other, neither Copy nor Merge is preferred at this point. If, however, Copy is chosen, then at the next step our choice is between merging the copy or merging *there*. Once again neither of these operations is obviously more economical than the other and so (1b) should be derivable. One way around this conclusion will not serve. One might be tempted to say that *overall* Merge is cheaper than Move, i.e. the derivation would terminate sooner were Merge chosen at this point than Move. However, as Chomsky noted in his original discussion, this is incorrect. For the derivation of (1b) and that of (2) involve the same number of Move and Merge operations overall, but only (2) is acceptable.

(2) There seems to be a man here

To distinguish (1b) from (2), it is critical that economy argument be *locally* evaluated (i.e. that we compare Merge versus Move at a given point in the derivation not overall) and for this we must assume that Move is a compiled operation.

There are various ways of motivating MOM consistent with this conclusion. One could, for example, propose that Merge is inherently cheaper than Copy. However, this involves finding a natural metric for valuing operations. What is nice about the proposal above is that it is based on the undeniable premise that if an operation B includes operation A as a proper subpart then B is more complex than A. This, however, tacitly assumes that primitive operations are compiled and that it is the compiled rules that are evaluated for economy.[9]

[9] A similar problem arises if Move is analyzed as Internal Merge. External Merge will involve the operations in (ia), Internal Merge those in (ib):

(i) a. Select α, select β, Merge α and β
b. Select α, select β within α, Merge α and β

It is plausible that search considerations make the second step of (ia) more "economical" than that of (ib). On the other hand, I can imagine rationalizations that lead to the opposite cost accounting. At any rate, it appears that both kinds of Merge involve three sub-operations and so more must be said to determine their relative prices.

Another option is to treat Move as AGREE plus (internal) Merge. However, for this to achieve the desired result we would have to assume that Move is the compilation of these two more basic operations. As such the logic of the situation remains the same.

To repeat, the above is not intended to endorse MOM as an economy condition, nor to suggest that there may not be other ways of motivating MOM reasoning. The point is simply to note that there are extant forms of reasoning that seem to tacitly assume that operations in particular grammars are compiled complexes of more primitive operations and whether this is indeed the case is an empirical matter. Of course, it raises potentially interesting research questions: how are complex operations compiled and are there principles for compiling them?

Chomsky has often noted that GB style theories contrast with earlier versions of generative grammar in not taking rules as construction specific.[10] For example, the Standard Theory contains rules like Passive and Raising. GB dispenses with such rules and replaces them with multiple applications of the simpler rule Move NP. When this rule applies in (3) we call the result "Raising." When it applies to (4) we call it "Passive."

(3) John_1 seems [t_1 to be here]

(4) John_1 was arrested t_1

Assuming this to be true, does it imply that rules like Raising and Passive are not part of the grammar? Or does it simply imply that they are not part of the basic architecture of FL? Put differently, what if anything prevents the conclusion that Passive and Raising exist in a speaker's language particular grammar as complex (compiled) rules, which include Move NP as a subpart. So far as I can tell, nothing of principle prevents this conclusion. Put positively, it is consistent with Chomsky's earlier arguments that construction-based rules are part of a native speaker's grammar even if they are not part of UG, viz. FL's basic architecture.

The same holds true for minimalist conceptions of UG. As noted above, it is possible that native speakers compile basic operations into more complex ones as part of the process of acquiring their particular grammars. The illustration above involves Move as the complex of Copy+Merge. Nothing prevents going further still and assuming that speakers have construction specific operations like Raising and Passive as part of their arsenal of rules. If this is correct, Minimalism *per se* is not incompatible with construction-based grammars, constructions simply being compiled complexes of basic operations.[11]

If this is even roughly correct, questions surrounding the "psychological reality" of grammars become quite subtle. There could be a sense in which a

[10] See Chomsky (1983).

[11] Cedric Boeckx (p.c.) notes another possibility: that FL contains complex operations like Merge and Move but these evolved from more basic operations evolutionarily. Thus, *now* FL contains only the compiled rules. But these have evolved from more primitive FLs that only contained the basic non-complex operations.

compiled operation is part of a person's linguistic competence (and so accurately describes a given speaker's cognitive grammatical state and so be part of a descriptively adequate characterization of the speaker's knowledge of language), without its being part of the basic structure of UG. This would contrast with the common GB conception in which a person's individual grammar simply is a valued version of the principles of UG, i.e. UG principles plus values for the open parameters. On this latter view "rules" are epiphenomenal; all that exists are the principles, their values and the effects of the interactions of these principles. As such, the form of the end state closely resembles that of the initial state; same principles, different values. This contrasts with the conception mooted above where the form of the grammatical rules that characterize the native speaker's competence do not exist in UG, though of course the sub-operations that have been compiled do.

Say that this is more or less correct. How might rules be compiled? At the simplest level, it is natural to think that compiling is driven by exposure. Here's the picture: the child's task is to develop a grammar for its language. It starts off with the basic operations and employs these to fit the incoming primary linguistic data (PLD). On the assumption that the PLD consists of <PF,LF> pairs, the following can illustrate the procedure. A child hears (5) and knows that the syntactic subject is the logical object of *kiss*.[12]

(5) John was kissed

UG allows the child to analyze this as follows: *John* is interpreted as the object of *kiss*. As all grammatical relations are discharged under Concatenate, there must be a relation between *John* and *kiss*. Thus there must be a copy of *John* concatenated with *kiss* and this must be labeled as the internal argument, i.e. V labels the constituent. Similarly, *John* must be concatenated with T for case. As case is discharged under agreement with a case licensing head, there must be concatenation between *John* and finite T. This provides the structure in (6).

(6) [$_T$ John [$_T$ T-finite [. . .[$_V$ kiss John]]]]

More structure is filled in based on further information. For example, that *kissed* is the past participle of *kiss* and the latter is transitive licenses the conclusion that there is a *v* projection. As movement is evidently required here, the *v* cannot be marking case, etc. This reasoning is based on UG principles plus a specific assumption concerning the inventory of basic operations. Given these, acquisition can be construed, in the first instance, as a curve fitting exercise with the PLD being the data set "fitted" by the basic operations and principles of UG. If it turns out that the same set of operations recur together frequently enough then it would be natural to box them as a unit to allow for more efficient

12 See Berwick (1980) for a detailed implementation of the outlined caricature.

use of grammatical resources. The compiling, for example, of Copy+Merge as Move would make a lot of sense if copying was generally followed by merging. Similarly for Concatenate+Label. The relevant point is that even if it is true that the basic operations are Concatenate, Copy, and Label there is no reason to deny that Move and Merge are part of native speaker's inventory of operations.

Nor need compiling stop with Merge and Move. Assume that the child encounters a sufficient number of passive sentences. Then it might be possible for the child to factor out a passive like rule that is sensitive to the morphology of the passive participle (i.e. *en*) and the thematic structure of the sentence. For example, if (6) is a robust data point in the PLD, then the child might use this information to develop a grammatical shortcut to generating the underlying derivation. If the child knows the form of passive participles and knows all that this implies (no case by *v*, thematic marking by V, concatenation to V and labeling by V) then this information can be stored together in one rule whose structural description and change can be stored as one complex operation. In this way construction specific rules of grammar can be compiled and can form the basis of a native speaker's competence. Note that on this conception, how "articulate" the rule is (i.e. how complex the structural description and/or structural change is) will be a function of how useful it is to compile this information which itself will be related to various performance factors, e.g. how common such "constructions" are. Presumably, the more common the construction, the greater the payoff to compiling the primitive operations into rules that are sensitive to particular contexts.[13] In short, it makes sense to think that compiling is costly but that it occurs when the payoff is frequent applicability in a wide range of cases.[14]

If this is correct, it suggests that the technology developed within earlier theories of generative grammar should be re-explored. In particular, prior to "Move α" the format for grammar rules involved SDs (structural descriptions) and SCs (structural changes). Great care was lavished on the relative cost of more versus less elaborate SDs and SCs. Questions like the following were central: What is the cost of context variables? What sorts of expressions can play the role of context variables? Are terminals possible context elements? Do all context variables have to be "affected" by the rule? Do they have to be "next" to something affected? And so forth.[15] These sorts of questions disappeared when Move α became the only "rule" of UG. However, as we have seen, there

[13] Observe that on this conception, learning consists of curve fitting with specific innately provided operations and restrictions plus the capacity to compile simple operations into more complex rules. The picture this suggests is different in spirit from the parameter setting model typical of GB-based accounts. See the next section for further discussion.

[14] This process of compiling complex rules need not be restricted to children. It is possible that adults also develop compiled rules.

[15] See Chomsky (1976) for an elaborate discussion.

being only a small number of UG operations does not preclude the possibility that these can be compiled and combined with contextual information to form "larger" more complex and more refined rules that native speakers regularly employ. Whether such compiling takes place is an open empirical question, even if one adopts the view that the basic operations of UG and FL are not construction specific.

If individual grammars involve rules like Move and allow complex SDs and SCs, it would be natural to return to the understanding of grammatical conditions in markedness terms rather than as absolute prohibitions. Recall that Ross (1967), like most work in early Generative Grammar, understood conditions as constraining the interpretation of variables. On this conception, it is possible to "violate" a condition if enough context is provided in the SD and SC, thereby eliminating or minimizing the variable. In this sort of system, no condition is inviolable, just costly to ignore. An illustration might help clarify this point. Consider the Movement Theory of Control (MTC) and let's stipulate (for purposes of illustration) that subject control over *promise* in examples like (7) violates minimality as *John* moves over *Mary*.

(7) John promised Mary [t to leave]

It has been argued that the acceptability of examples like (7) show that the MTC is false. However, that depends on what the rule is that relates *John* to its upper theta position. If it is Move (D/NP), then it violates minimality. Note that the SD and SC of this rule is (8) where "X" and "Y" designate variables.

(8) SD: X NP Y → SC: X NP Y
1 2 3 1 2 3

As the variables in (8) specify nothing, they can be safely ignored. However, explicitly noting their presence makes it clear how minimality could be understood as a markedness condition: NP movement involves variables (i.e. X in (8)) and as it is a constraint on variables that movement over variables obey minimality, application of (8) with X ranging over another c-commanding NP is prohibited. However, we can effect NP movement with another possible rule:

(9) SD X promise NP_1 NP_2 Y → X NP_2 promise NP_1 Y

Rule (9) is the movement rule for *promise with the context specified*. The movement in (9) is not over a variable and so minimality (understood as a constraint on variables) is irrelevant. Thus (9) is not blocked by the presence of a direct object, as (8) is. Of course, there is a cost to this as (9) is considerably more complex than (8). It involves at least two context variables, *promise* and NP_1. The complexity built into the rule reflects the supposition that (9) is more marked than (8) in the sense that evidence from the PLD is required to add it to the grammar alongside (8). In fact, this is what Rosenbaum (1967) argued originally; that rules like (9) are marked and that subject control in verbs like *promise* is marked.[16]

Let me end by making clear what is at issue. So far as I can see, two issues must be separated: what is the inventory of basic operations and are there compiled operations in the particular grammars of native speakers? One can, without betraying the letter or spirit of minimalism, answer "yes" to both questions. This would allow "constructions" a role in the study of grammar and it is possible that the technology from an earlier era will serve to theoretically structure investigations into how and when compiled operations are built.

7.2.2 *Language acquisition and parameter setting*

GB offers a principle and parameters vision of the structure of FL and this comes with a ready account of language acquisition. Language "learning" is a matter of setting the open parameters of underspecified principles and rules. A grammar for a language is a vector of settings/values for these open parameters, e.g. +*pro* drop, −V2, +Wh in situ, Move α (α = VP), head initial, etc. This has been a very influential account of language acquisition and it has generated considerable empirical and theoretical work of value. However, recently this view of things has fallen out of favor for two reasons. First, the empirical basis for parameters has been seriously challenged. Second, the epistemological utility of parameters has been questioned. Let's consider these two points in turn.

The strongest kind of argument for a parameter setting conception of language acquisition is that language learning (and change) comes in chunks. A small change in a single parameter can ramify through the grammar and result in a lot of apparent surface differences. This was (and is) a very attractive idea. However, as Newmeyer (2004, 2005) argues, the idea has not panned out as hoped. Empirically it has been hard to find grammatical phenomena that cluster around a single parametric value. More particularly, the bulk of the proposals to date suffer from the problem that what their proposed parameters cluster

16 As matter of fact, I believe that this is the wrong way to treat cases like (7). However, it is a perfectly reasonable proposal and it is simply false that recent innovations have made proposals like Rosenbaum's inadmissible. If, however, it is false to go the way of (9), as I currently believe it is, then this suggests that perhaps the power of complex rules is not actually available in grammar and this suggests that perhaps allowing grammars to write construction specific complex rules is not a good idea. However, this is not the place to consider this possibility. See Boeckx, Hornstein and Nunes (forthcoming) for discussion of what I take to currently be the correct analysis of cases like (7).

together appear to vary independently across languages. If so, then such clustering cannot be traced to a single parameter value.[17] This does not mean that parameter setting is not involved in language acquisition. But it does mean that the *best* kind of evidence for the theory of parameters has not been forthcoming. Newmeyer (2004) critically reviews the most promising proposals and finds them empirically wanting.[18] Similarly, Kayne in his recent work has argued against the existence of macro-parameters of the type advocated by Baker by demonstrating that the large-scale correlations one would expect to occur do not in fact obtain. In place of macro-parameters that tie together many kinds of changes, Kayne explores micro-parameters which can affect differences between languages singly. Newmeyer calls such parameters "rules" and rightly points out that they bleach the notion of parameters of much of its intrinsic interest.[19]

The second reason in favor of parameter setting models has been their ability to provide (at least in principle) an answer to Plato's Problem. The idea is that construing language acquisition as parameter setting eases the problem faced by the child for setting parameter values is easier than learning the myriad possible rules of one's native language. In other words, the PLD can be mined for parameter values more easily than it can be for rules. This too has proven to be less obvious on further consideration. For example, on a micro conception of parameters, the differences between grammars will be learned one by one, presumably on the basis of data that express them. Thus, in contrast to the GB vision in which parameters are set on the basis of one kind of linguistic data and other kinds follow along as free-riders, on the micro-parameter conception

[17] See Newmeyer (2004) for a good review of the major proposals. As he observes, the absence of such large scale parametric differences is "a cause for disappointment, not rejoicing (p. 209)." Newmeyer further notes (pp. 209–211) that the replacement of large scale parameters with micro-parameters tied to lexical variation amounts to little more than the observation that languages differ. These differences can be described parametrically but this has no obvious advantage over describing the differences in rule-based terms.

[18] Though I am no expert in these matters, his arguments look pretty good to me.

[19] One might reply that parameters still have the advantage of being binary so the range of variation in a parametric theory is more limited than one based on rules as the latter can differ arbitrarily. There are several responses to this: first, that parameters are binary is not inherent to a P&P account. This is an empirical claim in addition to the claim that settings are parametric. Second, it is not clear that the choices are parametric. This will depend on whether a given effect can be isolated to the presence of a single parameter and how wide this parameter's influence is. Given the large number of functional categories currently in play and the way that they can interact to produce a given surface phenomenon, it is not clear that one can reduce the presence of a given effect to the difference in the setting of one parameter. Last of all, if parameters are stated in the lexicon (the current view), then parametric differences reduce to whether a given language contains a certain lexical item or not. As the lexicon is quite open ended, even concerning functional items as a glance at current cartographic work makes clear, the range of variation between grammars/languages is also open ended. In this regard it is no different from a rule-based approach in that both countenance the possibility that there is no bound on the possible differences between languages.

there are no consequences from a parameter's being set one way or another. If this is so, parameter setting amounts to curve fitting to the PLD, just as rule learning would be.[20] It thus appears that looking for parameter values need not be much different than looking for patterns in linguistic data as a whole.

There is a second important issue: how much of the language must be surveyed to set a parameter? The relevant technical question concerning parameter learning turns out to be whether it is possible to set parameters independently of one another, i.e. once a parameter is fixed in value, the value will not change. If parameter values are independent, then the problem of parameter setting is considerably eased. However, it currently looks as if parameter values are intimately connected with values of each being (more or less) sensitive to values of all.[21] If this is correct, then setting parameters cannot proceed piecemeal and parameter values are only assigned considering the PLD *as a whole*. Thus, contrary to initial expectations, the existence of parameters need not localize the acquisition process. It would only do so if parameters were independent, which they appear not to be.

The upshot of both these conclusions has been to question the empirical and epistemological utility of parameter setting models of grammar. Minimalist considerations provide another route to a similar conclusion. GB postulates a finite number of binary parameters. These parameters are "internal" to UG in the sense that FL itself specifies them. Thus, according to GB, FL includes an enumeration of *possible* differences among grammars. It is for this reason that the number of possible grammars, though perhaps large, is finite.[22] Thus, natural language grammars can only differ from one another in finitely many ways. Given that FL is genetically determined, this means that the genome must specify both the invariant properties of NL grammars (i.e. the principles of UG) and the possible ways that these can be realized within a native speaker. The Minimalist Program follows GB in assuming that the invariants are specified genetically. In the account proposed here, these will be the inventory of basic operations plus the Path Minimality principle. However, it is less clear that a specification of the ways that grammars may differ is also part of the story. Nor is it clear that it should be. In fact, methodologically speaking, the burden of proof is on those that postulate UG specified parameters, as this is the richer theory. The same may be true biologically if we assume, as is conventional, that specifying information in the genome is costly and is only undertaken if the natural environment in which the genome operates cannot be counted upon to reliably supply the requisite information. If, however, the environment is reliable, then the relevant information need not be specified genetically and so

[20] This point is forcefully made in Newmeyer (2004, 2005).
[21] See Dresher (1999), Fodor (1998).
[22] Chomsky (e.g. 1982) considers this to be a discovery of some interest.

it will not be. In the case of language, the relevant question is whether the PLD is sufficient for the LAD to construct a grammar given just the invariant basic operations and principles for constructing them or whether building a grammar also requires the explicit endogenous specification of parametric options. As noted, to date, it is not clear what these parameters are nor whether specifying them actually aids the acquisition process.

Assume for a moment that the idea of specified parameters is abandoned. What then? One attractive property of the GB story was the picture that it came with. The LAD was analogized to a machine with open switches. Learning amounts to flipping the switches "on" or "off." A specific grammar is then just a vector of these switches in one of the two positions. Given this view there are at most 2^P grammars (P = number of parameters). There is, in short, a finite amount of possible variation among grammars.[23]

We can replace this picture of acquisition with another one. Say that FL provides the basic operations and conditions on their application (e.g. like minimality). The acquisition process can now be seen as a curve fitting exercise using these given operations. There is no upper bound on the ways that languages might differ though there are still some things that grammars cannot do. A possible analogy for this conception of grammar is the variety of geometrical figures that can be drawn using a straight edge and compass. There is no upper bound on the number of possible different figures. However, there are many figures that cannot be drawn (e.g. there will be no triangles with 20 degree angles). Similarly, languages may contain arbitrarily many different kinds of rules depending on the PLD they are trying to fit. However, none will involve binding relations in which antecedents are c-commanded by their anaphoric dependents or where questions are formed by lowering a Wh element to a lower CP.

Note that this view is not incompatible with languages differing from one another in various ways. Chapter 6 considered the possibility that all dependencies are formed by overt movement but with sometimes the upper copy and sometimes the lower copy retained. It is possible that some languages always follow one or the other convention. It is also possible that some mix and match, sometimes interpreting the bottom copy and sometimes the top.[24] Whatever the facts of the matter, nothing we have said here prevents languages from adopting broad policies of the former type. However, what is unexpected is that languages/grammars should divide neatly into one of two groups, those that exclusively pronounce the top copy and those that solely pronounce the bottom one. We expect, in other words, to find micro-variation (micro-parameters), i.e.

[23] Though P can be very large, say if the number of parameters is on the order of 50–60, a conservative estimate.

[24] Lidz and Idsardi (1998) suggest this conception of variation.

differences, and that these differences are ones that can be detected using PLD. Whether this is indeed what our research on variation has taught is something that I leave to the judgment of those more expert than myself.

7.2.3 *Islands*

The proposal outlined here has said nothing about one important set of GB generalizations; those relating to island effects. Parts of the theory of movement have been addressed: e.g. the requirement that movement be to c-commanding positions and most of the locality restrictions on A-movement, as these fall under either minimality or Greed. However, the present proposal says nothing substantive about the locality restrictions on A′-movement, viz. those that fall within the purview of Subjacency or the ECP.[25] It may be possible to extend the current proposals to cover some portion of the former, though the extension will depend heavily on assumptions concerning the structure of various islands. I would like to here quickly sketch one possible extension to certain island effects.[26] Island effects fall into two broad categories; weak islands and strong islands. The former include Wh Islands, Inner Islands and Neg Islands. The latter include Adjunct Islands, Complex Noun Phrase Islands and Subject Islands. I will ignore the former in what follows by assuming that they either fall under some version of minimality (as originally proposed in Rizzi 1990) or under some semantic condition (of the kind proposed by Szabolcsi and Zwarts 1993). Let's start with the first two strong islands; adjuncts and complex noun phrases. Adjunct islands and the relative clause version of complex noun phrase islands fall together as both involve extraction from adjuncts. Thus, whatever it is that prohibits movement out of adjuncts should extend to cases of movement out of relative clauses as the latter are just species of adjoined clauses.

(10) $X\hat{}[_{adjunct} \ldots Y \ldots]$

[25] A-movement is generally more restrictive than A′-movement and the observed locality effects fall under minimality or Greed. Thus, super-raising cases like (i) violate minimality while cases like (ii) violate some version of Greed.

(i) $John_1$ seems it was told t_1 that Bill left
(ii) $John_1$ seems t_1 is tall

[26] However, before proceeding I would like to make it clear that this extension is even more speculative than what has been attempted heretofore. It is more speculative in at least two ways. First, unlike the proposal outlined in Chapters 2–5, I have little empirical work to underwrite it. Second, the ideas have a more "technological" feel in that they revolve on substantive mechanics for how paths are determined. The first caveat makes these proposals less empirically grounded. The second renders it less theoretically interesting.

Both, then, will involve movement from complex expressions that are part of an unlabelled concatenate. Why might this kind of movement be grammatically prohibited? Consider the following possibility.[27]

Chapters 2 and 3 proposed that movement dependencies are regulated by minimality calculated in terms of paths. Shorter paths trump longer ones. What if we further assumed that all movement must be *on* a path and that paths are calculated in such a way that a non-labeled concatenate "breaks" a path. There are various ways of implementing this idea. The latter assumption is already implicit in slash category notations for category labels in GPSG accounts of movement. Information about an extraction site is passed up through slash categories and these link the movement's launch site and the overt position of the mover. Similarly, the g-projections in Kayne (1984) can be interpreted as a way of calculating paths. If unlabeled categories cannot g-project then movement from unlabeled constituents will violate connectedness. Or, if we assumed that a Mover must concatenate (adjoin) with every node between it and its target as proposed in Takahashi (1994), then once again movement will be blocked from the adjunct in (10). Y cannot concatenate with [X^Adjunct] as the latter is not labeled and so cannot be a concatenate.[28] Note that each of these implementations assumes that paths must be "continuous" in that there must be an unbroken series of labeled projections between "mover" and the target. Of the three implementations, the Takahashi one fits best with current proposals for it is clear why a missing label might make a difference; absence of a label prevents concatenation. However, the essential idea is basically the same in all three approaches.

The reasoning extends to noun complement constructions if we assume that sentential "complements" to nouns are actually adjuncts rather than

[27] See Boeckx (2008) for a similar approach to adjunct island phenomena based in part on Hornstein and Nunes (forthcoming).

[28] There are technical issues that must be clarified to make this work. Assume that if α dominates β and γ is adjoined to β then α dominates γ. Thus an adjunct will be dominated by every projection that dominates the expression it is adjoined to. Moreover, we must assume that the unlabeled concatenation of an adjunct and its target is maximal. Strictly speaking it is not dominated by a category with the same label (as it has no label). This will then require an expression taken from the adjunct to concatenate with it for there to be a continuous path. Last, we must resolve what to do with simple extraction *of* an adjunct as in (i):

(i) When/where/how did you play the piano

This can be "solved" technically by assuming that an expression that is an immediate daughter of a maximal projection is already locally related to that projection and so further adjunction is not required. In other words, Takahashi's conception is understood as stating that a Mover must be locally within the projection of every intervening maxP (i.e. α is locally contained in XP iff no YP intervenes). This is parallel to assuming that a Wh in Spec C need not adjoin to CP when moving out, as it is already in the local domain of C by being in its specifier. Similar reasoning will allow an adjunct immediately dominated by "^" to move from its adjunction position without first adjoining to "^".

complements, as proposed in Stowell (1981). Subjects can be similarly assimilated if, for some reason, they also break a path. Kayne (1984) achieves this result by conditioning g-projection by canonical head government. Specifically only constituents that are canonically head governed can g-project. In English, this requires being governed by a head on the left. As subjects are not so governed, they cannot g-project. Chomsky (1986a) achieves a similar result by prohibiting adjunction to thematically marked DPs. As subjects are theta-marked, adjunction is prohibited and this, plausibly, serves to break a path. However, the observant reader will have noticed neither of these proposals has a ready analogue given our current assumptions and this serves to separate subjects from adjuncts and complex noun phrases as regards islandhood. Whether this is a positive result is unclear. There appear to be languages in which the subject condition does not hold though adjuncts and complex noun phrases remain islands. Even in English, sentences like (11a) are more acceptable than those in (11b,c).

(11) a. What sorts of cars do you expect drivers of to carry high insurance
b. What sorts of cars did you meet people who drive
c. What sorts of cars did you get angry at Sue because people drive

At any rate, there seems to be a way of integrating *some* islands into the kind of general account outlined in the earlier chapters. What is less clear is whether this sort of approach is empirically adequate or theoretically sound.[29] Fortunately, these issues go beyond the scope of the present project and can be safely filed in the "future research" drawer.

7.2.4 *Reversibility*

One of the central architectural features of a grammar is that it is used to parse and produce sentences. How are grammars used to do this? Well, both parsing and production pair a "meaning" with a "sound" and as grammars produce PF-LF pairs, they should be useful in executing this task. Moreover, the way that grammars make PF-LF pairs available is by providing a finite recipe for generating them. An obvious question then arises: Are the operations that grammars use to "generate" PF-LF pairs analogous to those used to parse and produce sentences with PF-LF properties?[30] One plausible assumption

[29] Uriagereka (1999) treats subject islands as the result of spell out. This addition would be consistent with the present approach, though it does not follow from it. It hints however at another possibility; islands are interface phenomena having to do with the conversion of hierarchical structures to linear ones. This is proposed in Fox and Pesetsky (2005) and Hornstein, Lasnik and Uriagereka (2007). If correct, reducing islands to conditions of well-formed paths might be the wrong way to proceed.

[30] See Berwick and Weinberg (1984) and Phillips (1996) for discussion.

(though perhaps incorrect) is that more or less the same operations that generate sentences are used to parse them and produce them. In other words, there is a relatively transparent[31] relation between the primitives, principles, and operations of the grammar and those of the parser/producer.[32] A strong version of this thesis would be that all of the operations, principles, and constructs specified by the grammar are operations, principles, and constructs of the parser/producer as well.[33] Weaker versions would treat as transparent some properties of the grammar while opaquely construing others.

In light of the central role of grammars in parsing, I would like to propose that, *ceteris paribus*, more transparent grammars (i.e. grammars whose operations, principles, and constructs are more directly usable by the performance systems) should be preferred to less transparent ones. After all, if grammars are used, then it must be that linguistic structures are constructed in accordance with grammatical principles in real time. What better way to do this than to build these structures using the very same principles, operations and constructs that the grammar employs?[34] If this is correct, however, it has an interesting consequence: a grammar's operations and principles must be reversible. What I mean by this is that its operations and principles should be usable whether one is building a structure bottom-up (when deriving a sentence grammatically) or left to right (when one is parsing a sentence in real time). The direction of the "flow" of information should not affect the applicability of the principles. They should, in short, be directionally invariant (viz. reversible). Curiously, this property is less trivial than it might seem. Let me illustrate.

Consider the definition of n-subjacent in Chomsky (1986a: 30, (59)):

(12) β is n-subjacent to α iff there are fewer than $n + 1$ barriers for β that exclude α.

The definition in (12) is asymmetric, i.e. that β is n-subjacent to α does not imply that α is n-subjacent to β. In fact, α will always be 0-subjacent to anything it c-commands, as in such cases there can be no barriers for α that exclude β. (13) illustrates the point.

31 This term is borrowed form Berwick and Weinberg (1984).

32 I am going to use the terms "parser" and "producer" in the following discussion. However, what I intend is "parsing" and "producing." What is the relationship between "generating" grammatical objects and parsing and producing utterances? It is convenient to reify these processes and talk about parsers and producers. However, this should not be read as requiring the existence of these sorts of specialized objects. Rather, all that I am assuming is that whatever one does when parsing and producing take place, grammatical knowledge as characterized by grammars is involved. The question is does this place any interesting constraints on how to think about grammars.

33 Henceforth I will simply talk about the parser but what I suggest should be read as relating to the producer as well.

34 This would involve understanding the operations etc. procedurally, as recipes for how to build structure on line.

(13) $[_{CP}$ Who$_2$ did $[_{TP}$ you $[_{VP}$ meet $[_{DP}$ a man $[_{CP1}$ t$_2$ who$_1$ [t$_1$ likes t$_2$]]]]]]

In (13) CP and DP are barriers for t_2 as they are barriers that dominate t_1 and exclude *who*$_2$. However, there are no barriers that dominate *who*$_2$ but don't also dominate (hence exclude) t_2.

This illustrates what I mean by reversibility. A subjacency principle based on (12) will have a preferred direction of application (bottom-up), directionality is built into a system that adopts it.[35] This is not a criticism, just a fact. However, to the degree that reversibility is a desirable grammatical property, such principles would be disfavored.[36]

There are other clearly non-reversible operations and principles that have been proposed. Consider two more. The following "freezing principle" is often part of minimalist accounts:

(14) Freezing: A Case marked/checked DP is no longer subject to grammatical manipulation

This is generally understood as prohibiting a case checked DP from further merge/move operations. It is used to account for why Raising is prohibited from finite clauses (*John seems t is tall). Whatever the utility of Freezing,

[35] Note that earlier versions of subjacency did not have this asymmetry. Chomsky (1973: 247) had the following definition:

> . . . If Y is subjacent to X, either X and Y are contained in all the same cyclic categories . . . or they are in adjacent cycles.

This definition is clearly symmetrical. Similarly for Chomsky (1977: 73):

> I will understand the subjacency condition as holding that cyclic rule cannot move a phrase from position Y to position X (*or conversely*) [My emphasis, NH] in [(i)]:

(i) . . . X . . . $[_{\alpha}$ . . . $[_{\beta}$ . . . Y . . .] . . .] . . . x, where α,β are cyclic nodes.
For the present, I will take the cyclic nodes to be S and NP.

The change from a symmetrical to an asymmetrical definition is due to the change from an absolute conception of barrier/cyclic node to a relativized one. See next note for discussion.

[36] It is worth considering exactly how (12) leads to irreversibility. There are two parts: First, Chomsky (1986a) presents a relativized definition of Barrier (γ is a barrier *for* α, it is not a barrier *tout court*). Second, one computes distance between expressions in terms of exclusion, which makes good sense given the relativized definition. This contrasts with other accounts of subjacency where being a bounding node or a phase is an inherent property and distance is computed in terms of intervening nodes. Thus, for example, in (13) DP and CP are bounding nodes/phases for both t_2 and *Who*$_2$, while they are barriers for t_2 but not for *Who*$_2$. Furthermore, classical subjacency (which requires that there be at most one bounding node separating t_2 from *Who*$_2$) measures "distance between" in terms of nodes separating the two. Thus, subjacency is symmetric as the same number of nodes sits between t_2 and *Who*$_2$ regardless of whether one measures this top to bottom or bottom to top. As illustration consider the following definition of "between": The set of nodes between *Who*$_2$ and t_2 is the complement of the intersection of the nodes dominating *who*$_2$ and t_2 (i.e. the nodes dominating the target minus the nodes dominating the launch site viz. $-(\{CP\} \cap \{CP, TP, VP, DP, CP1\}) = \{TP, VP, DP, CP1\}$). In sum, reversibility fails to obtain in the Barriers conception because it has both a non-symmetric conception of barrier and a non-symmetric way of calculating distance.

it is not a reversible principle. This can be seen by considering the typical structure of an (A-)chain. The head of a DP chain is in a case position and the foot is in a theta position. That the head of such a chain is in a case position directly reflects (14), as Freezing prohibits further movement and so further extensions of the chain. However, consider how (14) applies in parsing, going left to right. If we assume that parsing builds a licit phrase marker, then it must proceed by moving a DP from its case position to, ultimately, its theta position (i.e. it will involve lowering, cf. note 35). However, if this is correct and we require grammars to be transparent then (14) does not hold when constructing the phrase marker left to right, for it would require moving a DP after it has been case checked into a theta position.[37]

Here is one last example, whose relevance should be apparent. Chapter 6 argued against AGREE and in favor of Move. More particularly, I argued that AGREE as encapsulated in the current Probe/Goal architecture of grammar has certain theoretical and empirical limitations. Interestingly, Probe/Goal systems are also not reversible. The reason is simple: built into the Probe/Goal system is the idea that Probes asymmetrically c-command their Goals. Movement is made up of two parts: an Agree relation holding between the probe and the goal, and movement of the goal to the probe that it agrees with (for EPP reasons of various sorts). What makes this irreversible, is that the Probe, which is the target of movement, must c-command Goal. This is fine for movement from theta to case positions as the latter c-command the former. However, it is not possible in the reverse situations as theta positions cannot "probe" elements in case positions as they do not c-command them. This prevents "movement" from case positions to theta positions in the left-to-right direction, dependent as it is on agreement holding. If this is correct, then AGREE-based systems of the Probe/Goal variety are not reversible.

Interestingly Move-based grammars are not similarly encumbered. Movement (being Copy and Merge) is reversible. Regardless of the "direction" in which the phrase marker is built, the operation is well defined. It is no harder to copy and merge "high" than it is to copy and merge low.[38]

What is the upshot of this? If correct, then one way of cataloguing grammatical proposals is by how reversible their operations are. Are there minimalist

[37] It is easy enough to replace (14) with another principle like (i) which is reversible.

(i) Once a DP's requirements are met, it is frozen.

A DP has two principal requirements. It must bear a theta role and it must be case marked. Or, we can assume, as in earlier accounts that case checking heads also bear a case feature that is checked when the DP case is checked. Under this assumption, the DP can raise to at most one case marked specifier. At any rate, there are alternatives to (14) that fit our requirements that are reversible. The point, however, is that there are also principles that do not.

[38] Just like with lowering rules in, for example, Chomsky (1977). See note 35.

reasons for favoring or valuing reversible grammars? Yes. Grammars interface with cognitive components that use grammars for parsing and production. Grammars that are reversible are easier to "use" for these purposes than ones that are not as the mapping from grammars to parsers is, in a sense, smoother, the higher the transparency of grammars to parsers. If one accepts, as minimalism often does, that smooth interface conditions are preferred (think of the conditions that Full Interpretation imposes on the outputs of the grammar), then a premium can reasonably be placed on transparent grammars. More interestingly, it appears that a reversibility condition imposes constraints on grammatical options ruling out many extant proposals in the grammatical literature. It also serves to more closely link performance considerations to grammatical ones and thereby loosens the dichotomy between competence and performance factors. This, in itself, is not a bad thing, as it potentially widens the domain of data relevant for grammatical evaluations. Of course, like all matters that are so abstract, the proof will be in the results.[39]

7.2.5 Features and clausal structure, or why there is movement

There is one final very large outstanding issue that I have touched on in footnotes and that it is worth making explicit here in the "future topics" section. Earlier chapters have provided an account of the basic operations of UG, segregating them into those that are part of the general cognitive architecture (viz. Concatenate, Copy (and check feature)) and those (one actually, viz. Label) that are specific to FL. The chapters outlined how, in concert, these basic operations would function to produce grammatical structures in accord with the laws of grammar as described in GB and related theories. If successful, this would amount to "deriving" the laws of grammar from assumptions that are cognitively and neurologically more primitive. It is for the reader to judge the degree of success. However, whatever the final judgment, there are several important properties of natural language grammars that this effort has not addressed. Perhaps the biggest is why grammars have movement at all.

Note that this question is different from a closely related (yet different) one: Is movement virtually conceptually necessary? This latter question has

[39] Two points: First, see Boeckx and Hornstein (2007) for the application of this kind of reasoning to the particular case of non-obligatory control. Second, the system developed here poses its own challenges with respect to transparency. For example, the interested reader might wish to consider how the Extension Condition (recall, Extension is derived from the fact that labeled expressions are grammatical atoms in Chapter 2) meshes with transparency. The short answer is "not well." There is a way of making Extension consistent with transparency when parsing is left to right. It requires progressive structural reanalysis as structure is built left to right. Depending on how the details of this are worked out, it may be possible to derive some of the features that Phillips (1996) does from grammars that build structure left to right. This, however, is a topic reserved for future discussion.

been addressed both here and in other minimalist work. Chomsky's current answer is that movement is just an instance of (the simplest) conception of Merge and that Merge must be part of any conceivable grammar. If he is right, then any grammar that has Merge will have Internal Merge (aka Move) as a special instance. The story outlined above also takes Move to be an inevitable by-product of the most basic operations Concatenate and Copy. Regardless of which of these proposals one endorses, the end point is the same: grammars are expected to have the *resources* for displacement/movement as part of their natural package of possible operations. In this sense, the existence of movement is not a surprise.

However, this does not by itself explain why natural language grammars contain long distance dependencies. They only note that such dependencies could easily be coded given FL's natural resources. It does not explain why the grammar contains non-local dependencies in the first place. The main reason for the presence of long distance dependencies is that grammars require that elements enter into diverse relations with multiple heads. For example, DPs must be theta-marked, case-checked, and (possibly) checked for an A′-feature like Wh *and* if these features must be checked against different heads then movement (a natural and inevitable part of UG given the basic operations of FL) can be used to get the DP close enough to the head to have it checked. Thus, the reason that grammars display movement is that expressions have sundry requirements that must be checked against *different* heads. Or to put this counterfactually: were we able to put multiple features on a single head (e.g. theta, case, and Wh features), then all of DP's requirements could be discharged on first merge. However, this is not possible, so movement ensues. The question is: why is this not possible? Why can't heads bear multiple sets of different kinds of features?[40] Nothing said so far addresses this question.

At present, two kinds of answers (guesses, really) have been advanced. The first is that movement is demanded by interface requirements.[41] One version of this is that language is used to convey information and this is greatly facilitated by moving constituents about. It is often noted that new (focus) or old (topic) information moves to the periphery of the clause. Perhaps this is because the Old/New information interface system "likes" to have things highlighted in some way and that moving the relevant material to the "edge" of the clause accomplishes this. If we further assume that the grammar is very obliging in that

[40] Observe that this is different from the question of why expressions have several kinds of features to check. Why doesn't a theta-role suffice? We discuss this a little below. Note, however, that it is not surprising that both theta-role and Wh/Focus/Topic features are available in grammars. This reflects the interests communicators have. What is more interesting is that these features have been imported into the grammar as formal movement licensing features. However, the problem discussed above obtains even if we hold the existence of such features constant. It is, thus, an additional concern.

[41] See Chomsky (1995a).

it facilitates the requirements of the interfaces (after all the system is perfect!), then movement to edges is what we might expect to find.

The second hunch is more formal. It relies on an inherent limit on how many different kinds of features a head can carry. The intuition goes back to Pollock (1989) and the explosion of Infl. The intuition behind this work is that there is a bi-uniqueness relation between heads and features. In other words, every head carries (more or less) one and only one set of features. For example, for Pollock, we distinguished Tense from Agreement features in Infl, each now contained in a separate T^0 and Agr^0 head. This intuition was carried forward into the minimalist program in various ways, though with some variation. The earliest approaches (e.g. Chomsky 1993, 1995a) segregated agreement and case information from theta-information, in distinction from GB theories where the two could be combined. For example, objects in GB were both assigned a theta-role and a case by V. In early minimalist proposals, V continued to theta-mark a DP but another head (either Agr or *v*) checked its case. The separation of case checking from theta-role assignment *for a given DP* extends to most contemporary accounts. Note that this is not exactly the Pollock intuition. A head can carry multiple features (e.g. *v* can both assign a theta-role and check case) but it cannot check both these features against the same DP. Nonetheless, the idea persists (and there is decent evidence in its favor) that for a given DP separate heads check these features.[42] Why this is so is conceptually unclear, however. Nonetheless, if such a separation of grammatical powers is an inherent property of how heads carry and check features, then movement would be required to get a head into the right local configurations for feature checking to take place.

Both these accounts have their attractions and their (obvious) drawbacks. Nor are they incompatible. However, they are both rather underdeveloped and rely on largely unknown factors. For example, we know relatively little about the interface components and so we can fill them with whatever requirements we need. Moreover, it is not at all clear why grammars should subserve interface requirements via movement. Take the Old/New information format. Couldn't the interface be served just as well with features that specify what's what? Some languages in fact do exactly this morphologically. Others use intonation and stress. Why is movement necessary? A possible riposte is that though not necessary, it is one way of marking these important distinctions and some

[42] The Chomsky (1993) variant was closer to Pollock's original intuition in that Agr(eement) heads were distinguished from Verbal heads. The latter were theta markers whereas the former checked case and agreement features. Chomsky (1995a) argued that Agr heads should be dispensed with as ϕ features in Agr projections are uninterpretable and so absent at LF. This led to sprinkling ϕ features onto heads like *v* and T^0 so that once again heads could carry complex feature sets. However, the vestige of the intuition remained in that a single head did not both assign a Θ role to and check the case of a given DP.

grammars take advantage of this opportunity, not surprisingly, as FL has the capacity to mark the distinctions in this way given the availability of movement. In fact, one could go further: if for some reason edges are salient and for some reason certain kinds of information are grammatically important, then *given that grammars can inherently move things around*, we would expect grammars to mark relevant distinctions by using the technology it naturally has, i.e. movement, in salience marking ways. The success of the idea would then depend on successfully defining the "edges" which would be salient. Given the kinds of movement we find, it would seem that *v*P, TP and CP mark salient edges. These do form a "natural" semantic class, all being propositional. And maybe these are the kinds of edges that informationally sensitive interfaces would recognize.

A particularly interesting version of this might advert to the observation that for semantic evaluation, natural language propositions (like Gaul) are divided into three parts; the nuclear scope, the restrictive clause and the quantificational scope. Grammar respects this three-way distinction by providing natural mapping rules from a phrase marker to the various sectors of the proposition. Thus, expressions within the *v*P map to the nuclear scope, those in the TP to the restrictive clause and those in the A′-domain to the quantificational scope, as proposed in Diesing (1992) developing ideas in Heim (1982). This kind of mapping hypothesis makes movement natural in that we might expect that a well-designed system would partition phrase markers in such a way as to respect propositional requirements. Thus, the reason that movement exists is because information relevant to the C-I interface is structured in propositional form and movement eases the mapping between sentences and propositions. Perhaps.

An equally congenial answer would focus on the limits inherent in packing different kinds of information together within a single head. It does appear empirically that it takes many separate projections to make a sentence. A plausible reason for this is that the way FL packages grammatical information requires the use of multiple heads. This makes sense if there is some limit to how much diverse kinds of grammatical information FL can pack into a given lexical item. If so, multiple heads are required and movement is called into action.[43]

[43] There still remain questions, however. For example, *why* does a DP need both a case feature and a theta-role? Why does an uninterpretable feature like case exist in a "perfect" system? After all, it appears to be a feature that exists only to be removed. It is the strangest feature of all, in this respect, and contrasts with Wh, Focus and Topic features, all of which have plausible interpretations and utility. Perhaps case serves to facilitate the mapping from the grammar to the tripartite proposition of the C-I interface, as Diesing (1992) suggests. Note, that this illustrates that the two kinds of answers can be combined: it is both true that heads have featural limits and that movement is for interface reasons.

The observant reader will have noted that though these ruminations may be natural, they are not definitive. Too little is known at present to go much beyond this kind of speculation. This acknowledged, I end here, hoping to have made it clear that this very important set of issues has hardly been addressed.[44]

7.3 A short philosophical postscript

Chapter 1 noted that the Minimalist Program (MP) is a continuation of the GB research program, not a competitor. It is a continuation in at least two ways. First, MP starts from the assumption that GB is roughly correct. It accepts both the general problems identified for solution (e.g. Plato's Problem) and the generalizations ("laws") that have been uncovered (at least to a good first approximation). The second way that MP continues the GB program is in its identification with the Rationalist research strategy that sits at the core of Chomskyan enterprise in general and GB in particular. As is well known, Rationalists, in contrast with Empiricists, endorse the postulate that the mind/brain is endogenously highly structured and that this structure channels experience in linguistically relevant ways. However, this is but a special case of the rationalist worldview. Rationalism's central characteristic is its general structuralist bias. This contrasts with the historical sensibility characteristic of empiricist approaches. Let me illustrate.

Consider how Rationalists and Empiricists differ concerning the sources of mental structure. The former think that mental structures arise largely through the operation of endogenous principles characteristic of the structure of the mind/brain. In the language case, these include the structure of FL as outlined by theories of UG. Empiricists, in contrast, believe that mental structures reflect exogenous factors, in particular the properties of environmental input whose operations on a relatively malleable mind serve to structure it. For Rationalists, the sources of cognition lie largely in the (innate) structures of the mind. For Empiricists, the sources of cognition lie in the processes through which the environment structures the mind.

This difference in attitude is not restricted to mental domains. The deeper difference is metaphysical. For Rationalists the observed world is the product of a small number of general interacting forces. For Rationalists, explanations are deeply deductive and phenomena are explained as special instances of these

44 Another possible motive for movement (and hence separate heads that forced it) would be to create structures able to support quantificational structure. It is interesting to note that movement of strong quantifiers to edge positions for reasons of case or various A′ features suffices to create structures capable of supporting quantificational interpretation. See Hornstein (1995) for one version of this without QR. This approach fits very well with Diesing's (1992) views on mapping to tripartite structure.

general principles. For Rationalists, initial conditions and historical contingencies matter less than underlying principles. In the main, Rationalists explain why X is Y by noting that X is of a type that has property Y: all things with structure X have property Y because Y is a necessary consequence of being something with the structure of X.

For Empiricists, in contrast, phenomena are not so tightly constrained by principle. Rather things hang together historically and contingently. Explanations are path dependent, responsive to the contingencies of history. Why is X Y? Because it arises from Y like things historically and because of X's history of development features of Y are preserved in X. Examples of this second mode of explanation are common: history of exposure to English results in a child's acquiring English, generations of bears in snowy backgrounds leads to white bears.

The above, of course, are caricatures. Rationalists can acknowledge the effects of historical happenstance and Empiricist theories do contain principles. However, it is also true, I believe, that the kinds of principles Empiricists find congenial are ones that leave a large role for exogenous (e.g. environmental) factors while Rationalists prefer accounts that take the effects of exogenous factors to be tightly constrained. Empiricists like path dependent accounts in which where one ends up is contingent on the path one takes, while Rationalists prefer principle dependent theories where the effects of initial conditions are washed quickly away by the forces endogenous to the system.

The Minimalist Program is a continuation of the rationalism of early generative grammar and roughly for the reasons it was pursued earlier. For example, Empiricist path dependent accounts need time to operate. To provide a historical explanation requires time. If language acquisition takes place quickly and uniformly regardless of the input, then it is unlikely that the output has the properties it does because of its history of exposure. Rather it has the properties it has because of the kind of thing it is, i.e. because of the structure it has. Similarly, if FL arose quickly then the structure it has cannot be a function of the details of the historical path that it took (as there was none) but because of options that were structurally available.

The proposal offered here is rationalist in this sense. The focus has been on what the background operations were and what sorts of FL would arise were Labeling added to the mix. It is in this sense, principle driven: given a structure with Concatenation, Copy and a principle minimizing dependency length, adding an operation like Label will result in an FL like GB. History be damned.

Early Generative Grammar had an impact far beyond linguistics precisely because it had consequences for this larger debate. There are times during which the largest research programs gain empirical traction, providing guidance and sustenance to one or another of the great philosophical approaches. This was so in the earliest days of Generative Grammar, and is arguably again so now.

In my opinion, what makes MP provocative, exciting, and fun is how it stems from these very large philosophical concerns and how it is able, at times, to generate hypotheses of empirical consequence. My aim here has been to explore the leading ideas of the Minimalist Program by proposing a theory compatible with (some) of its main tenets, which also has some empirical reach. The discussion has been conducted at a relatively abstract level, at least by the general standards in linguistics. Example sentences have been relatively rare and new proposals for particular paradigms or data sets have been wanting. The above, in short, has been largely a theoretical exercise. I believe that one of the more important consequences of the minimalist turn has been to make this sort of research both possible and possibly useful. I hope that the effort here spurs others to try their hands at this new game. Take it from me, it can be a lot of fun.

References

Abels, K. 2003. "Successive cyclicity, anti-locality, and adposition stranding," PhD thesis, University of Connecticut, Storrs.

Aoun, J. and D. Sportiche. 1983. "On the formal theory of government," *The Linguistic Review* 2: 211–236.

Avelar, J. 2004. "Dinâmicas Morofssintáticas com *Ter*, *Ser* e *Estar* em Português Brasileiro," MA thesis, Universidade Estadual de Campinas.

Baker, M. 1988. *Incorporation*. Chicago, IL: University of Chicago Press.

Belletti, A. 1988. "The case of unaccusatives," *Linguistic Inquiry* 19: 1–34.

Berwick, R. 1980. "Computational analogues of constraints on grammars: A model of syntactic acquisition," in *18th Annual Meeting of the Association for Computational Linguistics*.

Berwick, R. and A. Weinberg. 1984. *The grammatical basis of linguistic performance: Language use and acquisition*. Cambridge, MA: MIT Press.

Bhatt, R. 2005. "Long distance agreement in Hindi-Urdu," *Natural Language & Linguistic Theory* 23: 757–807.

Bobaljik, J. D. 1995a. "Morphosyntax: The syntax of verbal inflection," PhD thesis, MIT, Cambridge.

Bobaljik, J. D. 1995. "In terms of Merge: Copy and head movement," *MIT Working Papers in Linguistics* 27: 41–64.

Bobaljik, J. D. and S. Brown. 1997. "Interarboreal operations: Head movement and the extension requirement," *Linguistic Inquiry* 28: 345–356.

Bobaljik, J. D. and D. Jonas. 1996. "Subject positions and the roles of TP," *Linguistic Inquiry* 27: 195–236.

Bobaljik, J. D. and S. Wurmbrand. 2005. "The domain of agreement," *Natural Language & Linguistic Theory* 23: 803–865.

Boeckx, C. 2003a. *Islands and chains: Resumption as stranding*. Amsterdam: John Benjamins.

Boeckx, C. 2003b. "Intricacies of Icelandic agreement," ms., University of Maryland, College Park.

Boeckx, C. 2004. "Long-distance agreement in Hindi: Some theoretical implications," *Studia Linguistica* 58: 23–36.

Boeckx, C. 2006. *Linguistic Minimalism: Origins, concepts, methods, and aims*. Oxford: Oxford University Press.

Boeckx, C. 2008. *Bare syntax*. Oxford: Oxford University Press.

Boeckx, C. Forthcoming. "The nature of merge: Consequences for language, mind and biology," in M. Piatelli-Palmarini, J. Uriagereka and P. Salaburu (eds.), *Of minds and language: The Basque Country encounter with Noam Chomsky*. Oxford: Oxford University Press.

Boeckx, C. and N. Hornstein. 2004. "Movement under control," *Linguistic Inquiry* 35: 431–452.

Boeckx, C. and N. Hornstein. 2006. Raising and control. *Syntax* 9: 188–130.

Boeckx, C. and N. Hornstein. 2007. "On (non-)obligatory control," in W. D. Davies and S. Dubinsky (eds.), *New horizons in the analysis of control and raising*. Dordrecht: Springer, 251–262.

Boeckx, C., N. Hornstein and J. Nunes. 2007. "Overt copies in reflexive and control structures: A movement analysis," in A. Conroy, C. Jing, C. Nakao and E. Takahashi (eds.), *University of Maryland Working Papers in Linguistics* Vol. 15, 1–46.

Boeckx, C., N. Hornstein and J. Nunes. In Progress. *The movement theory of control.*

Boeckx, C. and Y. Jeong. 2004. "The fine structure of intervention in syntax," in C. Kwon and W. Lee (eds.), *Issues in current linguistic theory: A festschrift for Hong Bae Lee*. Seoul: Kyungchin, 83–116.

Boeckx, C. and S. Stjepanović. 2001. "Head-ing toward PF," *Linguistic Inquiry* 32: 345–355.

Bošković, Ž. 1999. "On multiple feature checking: Multiple wh-fronting and multiple head movement," in S. D. Epstein and N. Hornstein (eds.), *Working Minimalism*. Cambridge, MA: MIT Press, 159–187.

Bošković, Ž. 2007. "Agree, phases and intervention effects," *Linguistic Analysis* 33: 54–96.

Bowers, J. 2005. "On reducing control to movement," ms., Cornell University, Ithaca.

Cagri, I. 2005. "Minimality and Turkish relative clauses," PhD thesis, University of Maryland, College Park.

Castillo, J. C., J. Drury and K. K. Grohmann. 1999. "Merge over Move and the Extended Projection Principle," in S. Aoshima, J. Drury and T. Neuvonen (eds.), *University of Maryland Working Papers in Linguistics* Vol. 8, 63–103.

Chametzky, R. A. 1996. *A theory of phrase markers and the extended base*. Baffalo, NY: SUNY Press.

Chametzky, R. A. 2000. *Phrase structure: From GB to Minimalism*. Oxford: Blackwell.

Chandra, P. 2007. "(Dis)AGREE: Movement and agreement reconsidered," PhD thesis, University of Maryland, College Park.

Chomsky, N. 1955. "The logical structure of linguistic theory," ms., Harvard University and MIT. [Revised 1956 manuscript published in part as *The logical structure of linguistic theory* by New York, NY: Plenum, 1975; Chicago, IL: The University of Chicago Press, 1985.]

Chomsky, N. 1957. *Syntactic structures*. The Hague: Mouton.

Chomsky, N. 1959. "Review of B. F. Skinner *Verbal Behavior*," *Language* 35: 26–58.

Chomsky, N. 1964. *Current issues in linguistic theory*. The Hague: Mouton.

Chomsky, N. 1965. *Aspects of the theory of syntax*. Cambridge, MA: MIT Press.

Chomsky, N. 1976. "Conditions on rules of grammar," *Linguistic Analysis* 2: 303–351.

Chomsky, N. 1977. "On wh-movement," in P. W. Culicover, T. Wasow and A. Akmajian (eds.), *Formal syntax*. New York: Academic Press, 71–132.

Chomsky, N. 1981. *Lectures on government and binding*. Dordrecht: Foris.

Chomsky, N. 1983. "Some conceptual shifts in the study of language," in L. S. Cauman, I. Levi, C. D. Parson and R. Schwartz (eds.), *How many questions? Essays in honor of Sidney Morgenbesser*. Indianapolis, IN: Hackett.

Chomsky, N. 1986a. *Barriers*. Cambridge, MA: MIT Press.

Chomsky, N. 1986b. *Knowledge of language: Its nature, origin and use*. New York, NY: Praeger.

Chomsky, N. 1993. "A minimalist program for linguistic theory," in K. Hale and S. J. Keyser (eds.), *The view from Building 20: Essays in linguistics in honor of Sylvain Bromberger*. Cambridge, MA: MIT Press, 1–52.

Chomsky, N. 1995a. *The Minimalist Program*. Cambridge, MA: MIT Press.

Chomsky, N. 1995b. "Bare Phrase Structure," in G. Webelhuth (ed.), *Government and Binding theory and the Minimalist Program*. Oxford: Blackwell, 383–440.

Chomsky, N. 2000. "Minimalist inquiries: The framework," in R. Martin, D. Michaels and J. Uriagereka (eds.), *Step by step*. Cambridge, MA: MIT Press, 91–155.

Chomsky, N. 2001. "Derivation by phase," in M. Kenstowicz (ed.), *Ken Hale: A life in language*. Cambridge, MA: MIT Press, 1–52.

Chomsky, N. 2004. "Beyond explanatory adequacy," in A. Belletti (ed.), *Structures and beyond: The cartography of syntactic structures*. Oxford: Oxford University Press, 104–131.

Chomsky, N. 2005. "Three factors in language design," *Linguistic Inquiry* 36: 1–22.

Chomsky, N. 2005b. "On Phases," ms., MIT, Cambridge. [To appear in R. Freidin, C. P. Otero and M. L. Zubizarreta (eds.), *Foundational issues in linguistic theory*. Cambridge, MA: MIT Press.]

Chomsky, N. and H. Lasnik. 1993. "The theory of principles and parameters," in J. Jacobs, A. von Stechow, W. Sternefeld and T. Vennemann (eds.), *Syntax: An international handbook of contemporary research*. Berlin: Walter de Gruyter, 506–569.

Citko, B. 2003. "ATB wh-questions and the nature of Merge," in M. Kadowaki and S. Kawahara (eds.), *Proceedings of the thirty-third annual meeting of the North East Linguistic Society*. Amherst, MA: GLSA Publications.

Collins, C. 1997. *Local economy*. Cambridge, MA: MIT Press.

Collins, C. 2002. "Eliminating labels," in S. D. Epstein and T. D. Seely (eds.), *Derivation and explanation in the minimalist program*. Oxford: Blackwell, 42–64.

Cosmides, L. and J. Tooby. 1992. "Cognitive adaptations for social exchange," in J. Barkow, L. Cosmides and J. Tooby (eds.), *The adapted mind: Evolutionary psychology and the generation of culture*. Oxford: Oxford University Press.

Dawkins, R. 1996. *Climbing Mount Improbable*. Oxford: Oxford University Press.

de Marcken, C. 1996. "Unsupervised language acquisition," PhD thesis, MIT, Cambridge.

den Besten, H. 1983. "On the interaction of root transformations and lexical deletive rules," in W. Abraham (ed.), *On the formal syntax of the Westgermania*. Amsterdam: John Benjamins, 47–121.

den Dikken, M. 1995. "Binding, expletives and levels," *Linguistic Inquiry* 26: 347–354.

Diesing, M. 1992. *Indefinites*. Cambridge, MA: MIT Press.

Donati, C. 2003. "Merge copy," in K. Schwabe and S. Winkler (eds.), *The interfaces: Deriving and interpreting omitted structures*. Amsterdam: John Benjamins, 155–176.

Dresher, B. E. 1999. "Charting the learning path: Cues to parameter setting," *Linguistic Inquiry* 30: 27–67.

Embick, D. and D. Poeppel. 2005a. "Mapping syntax using imaging: Prospects and problems for the study of neurolinguistic computation," in K. Brown (ed.), *Encyclopedia of language and linguistics (2nd edn.)*. Oxford: Elsevier.

Embick, D. and D. Poeppel. 2005b. "Defining the relation between linguistics and neuroscience," in A. Cutler (ed.), *Twentiy-first century psycholinguistics: Four cornerstones*. Mahwah, NJ: Lawrence Erlbaum Associates, 103–118. [To be reprinted in A. Di Sciullo (ed.), *Biolinguistics*. MIT Press, 2008.]

Enç, M. 1991. "The semantics of specificity," *Linguistic Inquiry* 22: 1–25.

Epstein, S. D. 1999. "Un-principled syntax: The derivation of syntactic relations," in S. D. Epstein and N. Hornstein (eds.), *Working Minimalism*. Cambridge, MA: MIT Press, 317–345.

Epstein, S. D., E. M. Groat, R. Kawashima and H. Kitahara. 1998. *A derivational approach to syntactic relations*. Oxford: Oxford University Press.

Epstein, S. D. and T. D. Seely (eds.). 2002. *Derivation and explanation in the minimalist program*. Oxford: Blackwell.

Epstein, S. D. and T. D. Seely. 2006. *Derivations in Minimalism*. Cambridge: Cambridge University Press.

Fitzpatrick, J. 2006. "Deletion through movement," *Natural Language & Linguistic Theory* 24: 399–431.

Fodor, J. A. 1998. *In critical condition: Polemical essays on cognitive science and the philosophy of mind*. Cambridge, MA: MIT Press.

Fodor, J. A. 2000. *The mind doesn't work that way: The scope and limits of computational psychology*. Cambridge, MA: MIT Press.

Fodor, J. D. 1998. "Unambiguous triggers," *Linguistic Inquiry* 29: 1–36.

Fox, D. and D. Pesetsky. 2005. "Cyclic Linearization of syntactic structure," *Theoretical Linguistics* 31: 1–45.

Fujii, T. 2006. "Some theoretical issues in Japanese control," PhD thesis, University of Maryland, College Park.

Fukui, N. 1997. "Attract and the A-over-A principle," *UCI Working Papers in Linguistics* #3: 51–67.

Groat, E. M. 1999. "Raising the case of expletives," in S. D. Epstein and N. Hornstein (eds.), *Working Minimalism*. Cambridge, MA: MIT Press, 27–44.

Groat, E. M. and W. O'Neil. 1996. "Spell-out at the LF interface," in W. Abraham, S. D. Epstein, H. Thráinsson and C. J.-W. Zwart (eds.), *Minimal ideas*. Amsterdam: John Benjamins, 113–139.

Grohmann, K. K. 2003. *Prolific domains: On the anti-locality of movement dependencies*. Amsterdam: John Benjamins.

Gussenhoven, C. 1984. "Testing the Reality of focus domain," *Language and Speech* 26: 61–80.

Hale, K. and S. J. Keyser. 2002. *Prolegomenon to a theory of argument structure*. Cambridge, MA: MIT Press.

Harley, H. 2004. "Merge, conflation, and head movement: The first sister principle revisited," in K. Moulton and M. Wolf (ed.), *Proceedings of the thirty-fourth annual meeting of the North East Linguistic Society*. Amherst, MA: GLSA Publications.

Hauser, M. D., N. Chomsky and W. T. Fitch. 2002. "The faculty of language: What is it, who has it, and how did it evolve?," *Science* 298: 1569–1579.

Heim, I. 1982. "The semantics of definite and indefinite noun phrases," PhD thesis, University of Massachusetts, Amherst.

Heycock, C. 1994. *Layers of predication*. New York: Garland.

Hicks, G. 2006. "The derivation of anaphoric dependencies," PhD thesis, University of York, York.

Higginbotham, J. 1986. "Linguistic theory and Davidson's program in semantics," in E. Lepore (ed.), *Truth and interpretation: Perspectives on the philosophy of Donald Davidson*. Oxford: Basil Blackwell, 29–48.

Hornstein, N. 1995. *Logical form: from GB to Minimalism*. Oxford: Blackwell.

Hornstein, N. 1999. "Movement and control," *Linguistic Inquiry* 30: 69–96.

Hornstein, N. 2001. *Move! A minimalist theory of construal*. Oxford: Blackwell.

Hornstein, N. 2003. "On control," in R. Hendrick (ed.), *Minimalist Syntax*. Oxford: Blackwell, 6–81.

Hornstein, N. 2006. "Pronouns in a Minimalist setting," in N. Kazanina, U. Minai, P. J. Monahan and H. L. Taylor (eds.), *University of Maryland Working Papers in Linguistics*. College Park, MD: 47–80. Revised version printed in *The Copy Theory of Movement*. N. Corver and J. Nunes eds. 2007. Amsterdam/Philadelphia:John Benjamins.

Hornstein, N. 2007. "A very short note on existential constructions," *Linguistic Inquiry* 38: 410–411.

Hornstein, N. and H. Kiguchi. 2003. "PRO Gate and movement," in S. Arunachalam, E. Kaiser and A. Williams (eds.), *Proceedings of the 22nd Annual Penn Linguistics*.

Hornstein, N. and J. Nunes. 2002. "On asymmetries between parasitic gap and across-the-board constructions," *Syntax* 5: 26–54.

Hornstein, N., H. Lasnik and J. Uriagereka. (2003/out 2007). "The dynamics of islands: speculations on the locality of movement," *Linguistic Analysis* 33: 149–175.

Hornstein, N. and J. Nunes. 2008. "Some thoughts on adjunction," *Biolinguistics*.

Hornstein, N. and J. Uriagereka. 2002. "Reprojections," in S. D. Epstein and T. D. Seely (eds.), *Derivation and explanation in the minimalist program*. Oxford: Blackwell, 106–132.

Hornstein, N. and J. Witkos. 2003. "Yet another approach to existential constructions," in L.-O. Delsing, C. Falk, G. Josefsson and H. Sigurdsson (eds.), *Grammar in Focus Festschrift for Christer Platzack*. Lund: Department of Scandinavian Languages, University of Lund, 167–184.

Hunter, T. 2008. "The meaning of movement and adjunction," MS, University of Maryland, College Park.

Irurtzun, A. 2007. *The grammar of focus at the interfaces*. PhD Dissertation. The University of the Basque Country.

Jackendoff, R. 1994. *Patterns in the mind*. New York, NY: Basic Books.

Johnson, K. 1991. "Object positions," *Natural Language & Linguistic Theory* 9: 577–636.

Johnson, K. 2001. "What VP ellipsis can do, and what it can't, but not why," in M. R. Baltin and C. Collins (eds.), *The handbook of contemporary syntactic theory*. Oxford: Blackwell, 439–480.

Kato, Mary and Jairo Nunes. 1998. "Two sources for relative clause formation in Brazilian Portuguese," Paper presented at the Eighth Colloquium on Generative Grammar. Universidade de Lisboa.

Kawashima, R. and H. Kitahara. 1995. "Strict cyclicity, linear ordering, and derivational c-command," in J. Camacho, L. Choueiri and M. Watanabe (eds.), *The Proceedings of the Fourteenth West Coast Conference on Formal Linguistics*. Stanford, CA: CSLI Publications, 255–269.

Kayne, R. 1972. "Subject inversion in French interrogatives," in J. Casagrande and B. Saciuk (eds.), *Generative studies in Romance languages*. Rowley, MA: Newbury House, 70–126.

Kayne, R. S. 1984. *Connectedness and binary branching*. Dordrecht: Foris.

Kayne, R. 1994. *The antisymmetry of syntax*. Cambridge, MA: MIT Press.

Kayne, R. S. 2002. "Pronouns and their antecedents," in S. D. Epstein and T. D. Seely (eds.), *Derivation and explanation in the minimalist program*. Oxford: Blackwell, 133–166.

Kayne, R. S. 2005. "On parameters and on principles of pronunciation," in H. van Riemsijk and H. Broekhuis (eds.), *Organizating grammar: Studies in honor of Henk van Riemsdijk*. The Hague: Mouton de Gruyter, 289–299.

Kitahara, H. 1997. *Elementary operations and optimal derivations*. Cambridge, MA: MIT Press.

Kobele, G. 2006. "Generating copies," PhD thesis, UCLA, Los Angeles.

Koopman, H. 1984. *The syntax of verbs: From verb movement rules in the Kru languages to Universal Grammar*. Dordrecht: Foris.

Kuno, M. 2004. "Reconstruction obeys minimality: A representational theory of symmetrical grammar," *Linguistic Research: Working Papers in English Linguistics* 20: 39–101.

Lasnik, H. 1993. *Lectures on minimalist syntax*. Storrs: UConn Working Papers in Linguistics.

Lasnik, H. 1995. "Case and expletives revisited: On Greed and other human failings," *Linguistic Inquiry* 26: 615–633.

Lasnik, H. 1999. *Minimalist analysis*. Oxford: Blackwell.

Lasnik, H. 2001. "When can you save a structure by destroying it?," In M. Kim and U. Strauss (eds.), *Proceedings of the thirty-first annual meeting of the North East Linguistic Society*. Amherst, MA: GLSA Publications, 301–320.

Lasnik, H. and J. Kupin. 1977. "A restrictive theory of transformational grammar," *Theoretical Linguistics* 4: 173–196.

Lewis, R. L. and S. Vasishth. 2005. "An activation-based model of sentence processing as skilled memory retrieval," *Cognitive Science* 29: 375–419.

Lidz, J. and W. J. Idsardi. 1998. "Chains and phono-logical form," in A. Dimitriadis, H. Lee, C. Moisset and A. Williams (eds.), *Proceedings of the 22nd Annual Penn Linguistics Colloquium*. Philadelphia: Penn Working Papers in Linguistics, 109–125.

Lightfoot, D. 1990. "Old heads and new heads," in J. Mascaro and M. Nespor (eds.), *Grammar in progress: GLOW studies for Henk van Riemsdijk*. Dordrecht: Foris.

May, R. 1985. *Logical Form: Its structure and derivation*. Cambridge, MA: MIT Press.

McCloskey, J. 2000. "Quantifier float and wh-movement in an Irish English," *Linguistic Inquiry* 31: 57–84.

McKeown, R. 2007. "A movement account of Scandinavian reflexives," Generals Paper, University of Maryland, College Park.

Merchant, J. 2001. *The syntax of silence*. Oxford: Oxford University Press.

Müller, G. 1996. "A constraint on remnant movement," *Natural Language & Linguistic Theory* 14: 355–407.

Motomura, M. 2001. "Zibun as a residue of overt A-movement," Generals Paper, University of Maryland, College Park.

Murasugi, K. and M. Saito. 1995. "Adjunction and cyclicity," in R. Aranovich, W. Byrne, S. Preuss and M. Senturia (eds.), *The Proceedings of the Thirteenth West Coast Conference on Formal Linguistics*. Stanford, CA: CSLI Publications, 302–317.

Muysken, P. 1982. "Parameterizing the notion 'head'," *Journal of Linguistic Research* 2: 57–75.

Nakao, C. 2007. "Copy free movement, swiping and the ECP," Generals Paper, University of Maryland, College Park.

Newmeyer, F. J. 2004. "Typological evidence and Universal Grammar," *Studies in Language* 28: 527–548.

Newmeyer, F. J. 2005. *Possible and probable languages: A generative perspective on linguistic typology*. Oxford: Oxford University Press.

Nunes, J. 1995. "The copy theory of movement and linearization of chains in the Minimalist Program," PhD thesis, University of Maryland, College Park.

Nunes, J. 2001. "Sideward movement," *Linguistic Inquiry* 31: 303–344.

Nunes, J. 2004. *Linearization of chains and sideward movement*. Cambridge, MA: MIT Press.

Nunes, J. and Ž. Bošković. 2007. "The Copy Theory of Movement: A view from PF," in N. Corver and J. Nunes (eds.), *The Copy Theory of Movement*. Amsterdam: John Benjamins, 13–74.

Nunes, J. and J. Uriagereka. 2000. "Cyclicity and extraction domains," *Syntax* 3: 20–43.

Ortega-Santos, I. 2006. "On postverbal subjects, PF and the copy theory: The Spanish case," in N. Sagarra and A. J. Toribio (eds.), *Selected Proceedings of the 9th Hispanic Linguistics Symposium*. Somerville, MA: Cascadilla Press, 56–66.

Panagiotidis, P. 2003. "One, empty nouns and theta assignment," *Linguistic Inquiry* 34: 281–292.

Parsons, T. 1990. *Events in the semantics of English*. Cambridge, MA: MIT Press.

Pesetsky, D. 1982. "Paths and categories," PhD thesis, MIT, Cambridge.

Pesetsky, D. and E. Torrego. 2001. "T-to-C movement: Causes and consequences," in M. Kenstowicz (ed.), *Ken Hale: A life in language*. Cambridge, MA: MIT Press, 355–426.

Phillips, C. 1996. "Order and structure," PhD thesis, MIT, Cambridge.

Pietroski, P. 2005. *Events and semantic architecture*. Oxford: Oxford University Press.

Pietroski, P. 2007. "Induction and comparison," in A. Conroy, C. Jing, C. Nakao and E. Takahashi (eds.), *University of Maryland Working Papers in Linguistics*. College Park, MD: 154–186.

Pinker, S. 1997. *How the mind works*. New York, NY: W. W. Norton & Company.

Pires, A. 2006. *The minimalist syntax of defective domains*. Amsterdam: John Benjamins.

Poeppel, D. and P. J. Monahan. In press. "Speech perception: Cognitive foundations and cortical implementation," *Current Directions in Psychological Science*.

Polinsky, M. and E. Potsdam. 2002. "Backward control," *Linguistic Inquiry* 33: 245–282.

Polinsky, M. and E. Potsdam. 2006. "Expanding the scope of control and raising," *Syntax* 9: 171–192.

Pollock, J.-Y. 1989. "Verb movement, Universal Grammar, and the structure of IP," *Linguistic Inquiry* 20: 365–424.

Reinhart, T. 1983. "Coreference and bound anaphora: A restatement of the anaphora questions," *Linguistics and Philosophy* 6: 47–88.

Richards, N. 2001. *Movement in language: Interactions and architectures*. Oxford: Oxford University Press.

Rizzi, L. 1990. *Relativized minimality*. Cambridge, MA: MIT Press.

Rizzi, L. 2005. "Phase theory and the priviledge of the root," in H. van Riemsijk and H. Broekhuis (eds.), *Organizating grammar: Studies in honor of Henk van Riemsdijk*. The Hague: Mouton de Gruyter, 529–537.

Roberts, I. 1994. "Two types of head movement in Romance," in N. Hornstein and D. Lightfoot (ed.), *Verb movement*. Cambridge: Cambridge University Press, 207–242.

Rodrigues, C. 2004. "Impoverished morphology and movement out of case domains," PhD thesis, University of Maryland, College Park.

Rosenbaum, P. S. 1967. *The grammar of English predicate complement constructions*. Cambridge, MA: MIT Press.

Ross, J. R. 1967. "Constraints on variables in syntax," PhD thesis, MIT, Cambridge.

Saito, M. 1989. "Scrambling as semantically vacuous A′-movement," in M. R. Baltin and A. S. Kroch (eds.), *Alternative concepts of phrase structure*. Chicago: The University of Chicago Press, 182–200.

Saito, M. and N. Fukui. 1998. "Order in phrase structure and movement," *Linguistic Inquiry* 29: 439–474.

Samuels, B. 2008. "A string theory of syllables," MS, Harvard University.

Schein, B. 1993. *Plurals and events*. Cambridge, MA: MIT Press.

Seely, T. D. 2006. "Merge, derivational c-command, and subcategorization in a label-free syntax," in C. Boeckx (ed.), *Minimalist essays*. Amsterdam: John Benjamins, 182–217.

Selkirk, E. 1984. *The relationship between sound and structure*. Cambridge, MA: MIT.

Soltan. U. 2007. "On formal feature licensing in minimalism," PhD thesis, University of Maryland, College Park.

Sportiche, D. 1988. "A theory of floating quantifiers and its corollaries for constituent structure," *Linguistic Inquiry* 19: 425–449.

Stabler, E. 2007. "Language structure, depth and processing." Paper presented Mayfest 2007. University of Maryland, College Park.

Starke, M. 2001. "Move dissolves into merge: A theory of locality," PhD thesis, University of Geneva, Geneva.

Stjepanović, S. 2003. "A word-order paradox resolved by copy deletion at PF," *Linguistic Variation Yearbook* 3: 139–177.

Stowell, T. A. 1981. "Origins of phrase structure," PhD thesis, MIT, Cambridge.

Szabolcsi, A. and F. Zwarts. 1993. "Weak islands and an algebraic semantics of scope taking," *Natural Language Semantics* 1: 235–284.

Szczegielniak, A. 2005. "VP ellipsis and topicalization," in L. Bateman and C. Ussery (eds.), *Proceedings of the thirty-fifth annual meeting of the North East Linguistic Society*. Amherst, MA: GLSA Publications, 603–614.

Takahashi, D. 1994. "Minimality of movement," PhD thesis, University of Connecticut, Storrs.

Tallerman, M. 2006. "Challenging the syllabic model of 'syntax-as-it-is'," *Lingua* 116: 689–709.

Taraldsen, K. T. 1996. "Reflexives, pronouns and subject/verb agreement in Icelandic and Faroese," in J. R. Black and V. Motopanyane (eds.), *Microparametric syntax and dialect variation*. Amsterdam: John Benjamins, 189–211.

Uriagereka, J. 1995. "Aspects of the syntax of clitic placement in Western Romance," *Linguistic Inquiry* 26: 79–123.

Uriagereka, J. 1998. *Rhyme and reason: An introduction to minimalist syntax*. Cambridge, MA: MIT Press.

Uriagereka, J. 1999. "Multiple Spell-Out," in S. D. Epstein and N. Hornstein (eds.), *Working Minimalism*. Cambridge, MA: MIT Press, 251–282.

Uriagereka, J. 2002. "Pure adjuncts," ms., University of Maryland, College Park.

Vassilieva, M. and R. Larson. 2001. "The semantics of the plural pronoun construction," in R. Hastings, B. Jackson and Z. Zvolenszky (eds.), *Proceedings of Semantics and Linguistic Theory (SALT) XI*. Ithaca: CLC Publications, Department of Linguistics, Cornell University.

Vikner, S. 1991. "Verb movement and the licensing of NP positions in the Germanic languages," PhD thesis, University of Geneva, Geneva.

Wagner, M. 2005. "Prosody and recursion," PhD thesis, MIT, Cambridge.

Wilder, C. 1999. "Right node raising and the LCA," in S. Bird, A. Carnie, J. D. Haugen and P. Norquest (eds.), *Proceedings of the Eighteenth West Coast Conference on Formal Linguistics*. Somerville, MA: Cascadilla Press, 586–598.

Yoshida, M. 2006. "Sometimes smaller is better: sluicing, gapping and semantic identity," in C. Davis, A. R. Deal and Y. Zabbal (eds.), *Proceedings of the thirty-sixth annual meeting of the North East Linguistic Society*. Amherst, MA: GLSA Publications, 681–694.

Zwart, C. J.-W. 2002. "Issues relating to a derivational theory of binding," in S. D. Epstein and T. D. Seely (eds.), *Derivation and explanation in the Minimalist Program*. Oxford: Blackwell, 269–304.

Index